HISTORICAL GEOLOGY
LAB MANUAL

HISTORICAL GEOLOGY
LAB MANUAL

Pamela J. W. Gore
Georgia Perimeter College

WILEY

VP & PUBLISHER:	Petra Recter
EXECUTIVE EDITOR:	Ryan Flahive
ASSISTANT EDITOR:	Julia Nollen
EDITORIAL ASSISTANT:	Kathryn Hancox
MARKETING MANAGER:	Suzanne Bochet
PHOTO EDITOR:	Elizabeth Blomster
DESIGNER:	Kenji Ngieng
ASSOCIATE PRODUCTION MANAGER:	Joyce Poh

This book was set by Laserwords Private Limited. Cover and text printed and bound by Quad/Graphics.

This book is printed on acid free paper.

Founded in 1807, John Wiley & Sons, Inc. has been a valued source of knowledge and understanding for more than 200 years, helping people around the world meet their needs and fulfill their aspirations. Our company is built on a foundation of principles that include responsibility to the communities we serve and where we live and work. In 2008, we launched a Corporate Citizenship Initiative, a global effort to address the environmental, social, economic, and ethical challenges we face in our business. Among the issues we are addressing are carbon impact, paper specifications and procurement, ethical conduct within our business and among our vendors, and community and charitable support. For more information, please visit our website: www.wiley.com/go/citizenship.

ISBN 978-1-118-05752-0 (paperback)

Printed in the United States of America

V10018727_052120

TABLE OF CONTENTS

Preface

This laboratory manual is designed for introductory geology students with no prior geology coursework. The chapter on Relative Dating gives the student experience with using basic geologic principles for determining the sequence of geologic events, a topic that is typically presented in the first few chapters of historical geology lecture textbooks. The chapter on Rocks and Minerals provides a quick introduction to the minerals and rocks most commonly encountered in a first geology course, as ell as a quick review for students who have previously completed physical geology lab. Subsequent labs deal with Rock Weathering and Interpretation of Sediments, Sedimentary Rocks, and Sedimentary Structures. Students learn to interpret processes acting in the depositional basin, and then they interpret Depositional Sedimentary Environments based on rock types and sedimentary structures. Students also learn Stratigraphy and Lithologic Correlation and how to interpret sea-level change in the rock record.

The last five labs deal with fossils and include an overview of the latest classification system. These labs cover Microfossils and Introduction to the Tree of Life, Invertebrate Macrofossils and Classification of Organisms, Fossil Preservation and Trace Fossils, and two web quests: Fossils on the Internet, and The Evolution of the Vertebrates.

This laboratory manual has evolved over time from labs that I began writing as a graduate student at George Washington University in the early 1980s. I was teaching Historical Geology labs and did not find that any of the lab manuals fit my style of teaching and goals for the course. I wanted students to have a strong hands-on experience with geologic specimens and I wanted them to have fill-in charts where they could record their observations and interpretations on pages that they could submit for grading without having answers interspersed throughout the lab. The lab manual needed to be accessible to nonscience majors and also provide a strong background for Geology majors and other science majors. It is important for students to learn that they can think and make interpretations, and not just memorize vocabulary. I wanted concepts that carried over from one lab to the next, and labs that reinforced and built on previous labs, so that at the end of the semester, the students would have some experience at interpreting the rock record and some understanding of how the process of science works. I extend my appreciation to the students and faculty at George Washington University in the early 1980s (including Roy Lindholm, Tony Coates, John Lewis, David Govoni, and the late George Stephens) for inspiring and encouraging my early efforts.

I revised and updated the labs many times over the years while a Professor of Geology at Georgia Perimeter College (GPC), receiving comments, criticisms, and suggestions from students and Geology faculty colleagues (including Lynn Zeigler, Polly Bouker, Deniz Ballero, Kimberly Schulte, Gerald Pollack, Dion Stewart, John Anderson, Ed Albin, Rick Nixon, Debia McCulloch and others), as well as colleagues at other institutions in Georgia and across the country. In 1998, former student Gwendolyn Rhodes scanned my original artwork to help me get the lab manual online. The lab manual was available online for a number of years, and colleagues at colleges, universities, and schools around the world used all or parts of it with their classes.

As a result, I received emails from a worldwide audience including students and faculty in places such as Mauritius, Japan, Egypt, Morocco, Indonesia, Chile, Saudi Arabia, the Philippines, Oman, Iran, Iraq, Hong Kong, Belgium, Scotland, Ireland, England, Wales, Canada, Vietnam, Thailand, Brazil, Nigeria, Pakistan, Malaysia, and most of the states in the U.S. For a number of years, an early version of the lab manual was produced by the GPC Printshop in black and white, 3-hole punched. Some color pages were introduced in 2009, and students quickly demanded that the entire lab manual be made available in color. Beginning in 2010, the GPC Printshop produced the manual in full color with spiral-binding thanks to Barbara Lindsay Gatewood, Beverly Kelly, and others. I sincerely thank everyone who has contributed comments and suggestions to this lab manual, and everyone who has used it in their classes.

I wish to thank numerous colleagues who have contributed superb photographs to this edition, and museum staff that allowed photography or provided photographs, which have enhanced the laboratory manual considerably.

Thanks are also offered to Richard Hightower of The Stones and Bones Collection, and to Henry Crowley & Terry Lee of H & T Fossils, and others, for allowing me to photograph some of their amazing specimens.

Special thanks are extended to those at John Wiley & Sons, Inc. who have made this lab manual possible, including Ryan Flahive, Kathryn Hancox, Elizabeth Blomster, Joyce Poh, Julia Nollen, and many others.

And finally, thanks are extended to my family—my husband, Thomas J. Gore III, and daughter, Miranda J. Gore, who accompanied me to many of the localities photographed in this book. In addition, thanks are extended to Miranda for providing several of her photographs to this manual.

GUIDELINES FOR USING ACID TO
IDENTIFY MINERALS

CAUTION!
BE VERY CAREFUL
if you use acid in the lab

The hydrochloric acid used in geology lab is typically dilute, but if handled improperly it can be hazardous. Hydrochloric acid can cause acid burns to the skin or eyes, and it will burn holes through clothing.

Whenever using acid, *always*:

- **Follow directions carefully.**

- **Use a dropper bottle, and only apply one or two small drops to the sample.** Examine the sample to look for tiny gas bubbles. If the mineral is calcite, you will see them right away.

- Use care so you do not get the acid in your eyes or on the lab table, on your skin or clothes, or on any other person.

- Do not have any food or drink in the lab when acid is in use.

- Rinse specimens with tap water and blot with a paper towel after performing acid tests. Do not leave wet or acid-covered specimens on the lab tables, and do not put wet or acid-covered specimens back into the specimen trays.

- If you get acid on your hands, wash your hands immediately and notify your instructor.

- If you notice any adverse reactions after washing your hands thoroughly, get prompt medical attention.

- Do not rub your eyes after doing an acid test until after you have washed your hands.

- If acid get in your eyes, immediately flush your eyes with water for a minimum of 15 minutes using the emergency eyewash. Get prompt medical attention.

- If acid is spilled on other areas of the body, inform the instructor and use the safety shower, as appropriate. If the affected area is underneath clothing, remove the clothing. Get prompt medical attention.

- Report any acid spill immediately to the instructor.

- Close acid bottles securely, and return them to the acid storage area.

Follow your instructor's directions.

PROPER USE OF THE HAND LENS TO STUDY GEOLOGIC SAMPLES

1. Start with a 10× hand lens, and swing it open.

2. Curl your first finger and put it through the opening in the hand lens cover.

3. Hold the hand lens cover between your thumb and first finger.

4. Bring your hand up to your face so that your thumb is resting against your cheek, so that you can see through the hand lens.

5. Using your other hand, bring your geologic sample up toward your face, looking at it through the hand lens, until it is in focus, which will be about a half inch from your face.

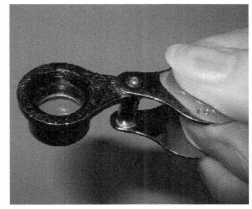

Pamela Gore

Pamela Gore

Relative Dating

This lab introduces the concept of **relative dating** of geologic sequences. Relative dating means determining which rock units are older and which are younger in a particular geologic setting. **Stratification** (layering or bedding) is the most obvious large-scale feature of sedimentary rocks (Figure 1.1). Bedding is readily seen in a view of the Grand Canyon or in almost any other sequence of sedimentary rocks. Each of the beds or **strata** is the result of a natural event in geologic history, such as a flood or storm. As time passes, many such events occur and the sediment piles up layer upon layer. In this way, thick sedimentary sequences are formed.

Grand Canyon National Park

Figure 1.1 Example of stratification or beds of sedimentary rock in the Grand Canyon, Arizona. View from Pima Point on the West Rim Drive, Grand Canyon National Park.

BASIC PRINCIPLES OF GEOLOGY

Steno's Laws

In a sedimentary sequence *the older beds are on the bottom and the younger beds are on the top*. You can visualize how this occurs if you imagine a stack of newspapers in the corner of a room. Every day you put another newspaper on the pile. After several weeks have passed, you have a considerable stack of newspapers, with the oldest ones on the bottom of the pile and the most recent ones on the top (Figure 1.2).

This fairly obvious but very important fact about layering—that older layers are on the bottom and younger ones are on top—was first noted in the 1600s by the Danish geologist Nicholas Steno. The **principle of superposition**, as it has come to be called, is the first of three principles now known as **Steno's laws** (Box 1.1). These three principles are based on observation and logic.

Pamela Gore

Figure 1.2 Vertical geologic section in the Fountain Formation. Layer A is the oldest, layer D is the youngest. Garden of the Gods, Colorado Springs, Colorado.

Steno's second law is the **principle of original horizontality**, which states that *sediments are deposited in flat, horizontal layers*. We can recognize this easily if we consider a sedimentary environment such as the sea floor or the bottom of a lake. Any storm or flood bringing sediment to these environments deposits it in a flat layer on the bottom because sedimentary particles settle under the influence of gravity. As a result, a flat, horizontal layer of sediment is deposited.

Steno's third law is the **principle of original lateral continuity**. If we consider again the sediment being deposited on the seafloor, *the sediment is not only deposited in a flat layer, it is a layer that extends for a considerable distance in all directions*. In other words, the layer is laterally continuous. Steno's laws are listed in Box 1.1.

Box 1.1 Steno's Laws
Principle of superposition
Principle of original horizontality
Principle of original lateral continuity

Of the three, the principle of superposition is most directly applicable to relative dating. We can examine any sequence of sedimentary strata and determine in a relative sense which beds are older and which beds are younger. All we need to know is whether the beds are right-side up or not.

This complication comes because tectonic forces can cause sedimentary sequences to be tilted, folded, faulted, and overturned (Figure 1.3). Although sediments are originally deposited in horizontal layers, they do not always remain horizontal. A trip to the mountains or a quick look through your textbook should convince you that many sedimentary sequences consist of layers or beds that dip at some angle to the horizontal, and in some cases, the beds are vertical or overturned.

Figure 1.3 Overturned fold in the Rockmart Slate, Rockmart, Georgia.

Lithologic Symbols

In diagrams, geologists use a standard set of **lithologic symbols** (from the Greek *lithos*, meaning "stone") to show rock types in diagrams such as stratigraphic sections or geologic cross sections. These symbols are used throughout this lab manual, and you will also find them used in your lecture textbook. Take a look through your textbook and see how many of the symbols in Table 1.1 you can identify in the geologic cross sections.

TABLE 1.1 LITHOLOGIC SYMBOLS

Symbol	Lithology	Symbol	Lithology
	Breccia		Limestone
	Conglomerate		Dolostone
	Sandstone		Volcanic rocks
	Siltstone		Plutonic rocks
	Shale		Plutonic rocks
	Coal		Metamorphic rocks

Other Basic Principles

In addition to Steno's laws, a number of other basic geologic principles can be used for relative dating.

The Principle of Intrusive Relationships

Where an igneous intrusion cuts across a sequence of sedimentary rock, the relative ages of these two units can be determined. *The sedimentary rocks are older than the igneous rock that intrudes them.* In other words, the sedimentary rocks had to be there first, so that the igneous rocks would have something to intrude. You could also say *the intrusion is younger than the rocks it cuts.* Examples of types of igneous intrusions (or **plutons**) are dikes, sills, stocks, and batholiths. **Dikes** are relatively narrow tabular intrusions that cut across the layers in the sedimentary rocks (Figure 1.4). **Sills** have intruded along the layers in the sedimentary rocks. Stocks and batholiths are larger, irregular shaped plutons. Batholiths are the larger of the two, by definition, covering more than 100 km^2 (or about 40 mi^2), whereas stocks are less than 100 km^2. The principle of intrusive relationships is illustrated in Table 1.2 using three types of igneous intrusions.

TABLE 1.2   PRINCIPLE OF INTRUSIVE RELATIONSHIPS

Illustration	Pluton type	Relative dating of layers
	Dike	Dike B is younger than sedimentary rock sequence A. Erosion surface C is younger than dike B. Sedimentary rock D is younger than erosion surface C.
	Dike and sill	Sill B is younger than sedimentary rock sequence A. Dike C is younger than sill B.
	Stock and dike	Stock B is younger than sedimentary rock sequence A. Dike C is the youngest.

Pamela Gore

Figure 1.4 Jurassic-age diabase dike (*black*) cutting through the Norcross Gneiss. Vulcan Materials Quarry, Norcross, Georgia.

The Principle of Cross-Cutting Relationships

A **fault** is a crack in the rock along which movement has occurred (Table 1.3). Where a fault cuts across a sequence of sedimentary rock, the relative ages of the fault and the sedimentary sequence can be determined. *The fault is younger than the rocks it cuts.* The sedimentary rocks are older than the fault that cuts them, because they had to be there first before they could be faulted.

When observing a faulted sequence of sedimentary strata, always look to see how the beds on either side of the fault have been displaced. You might be able to locate a "key bed" that has been offset by the fault. If so, you will be able to determine the type of fault. Two types of faults are discussed in this lab: normal faults and reverse faults.

In a **normal fault**, the hanging wall (HW), or the block of rock physically *above* the fault plane, moves down with respect to the foot wall (FW). Normal faults occur in response to tensional stress. Normal faults tend to occur at or near divergent tectonic plate boundaries (Figure 1.5).

In a **reverse fault**, the hanging wall, or the block of rock physically above the fault plane, moves up with respect to the foot wall. Reverse faults occur in response to compressional stress. Reverse faults tend to occur at or near convergent tectonic plate boundaries. Thrust faults are a type of low-angle reverse fault (Figure 1.6).

Principle of Components or Inclusions

In a sequence of sedimentary rocks, if there is a bed of gravel, *the clasts (or inclusions) of gravel are older than the bed in which they are contained.*

In many instances, the gravel directly overlies an irregular **erosion surface** (Figure 1.7). Sometimes it is obvious from the lithology that the clasts in the gravel bed are derived from the underlying partially eroded layer. If this is the situation, it is possible to place several layers and events in their proper relative order: (1) deposition of sedimentary rock sequence A; (2) erosion of sedimentary rock sequence A, producing an irregular erosional surface and rip-up clasts; (3) deposition of rip-up clasts of sedimentary rock A on top of the irregular erosional surface, producing a gravel bed. This gravel bed is sometimes called a **basal conglomerate** because it is at the base of the sedimentary sequence overlying the erosional surface.

A similar line of reasoning may be applied to igneous rocks if **xenoliths** are present (Figure 1.8). You may remember from physical geology that a xenolith (which literally means "foreign rock") is a piece of surrounding rock (sometimes called "country rock") which becomes caught up in an intrusion. As magma moves upward, forcing itself through cracks in the surrounding rock, sometimes pieces of these surrounding rocks break off or become dislodged and incorporated into the magma without melting. These pieces of rock are called *xenoliths*, and they move along with the magma. According to the principle of components or inclusions, *xenoliths are older than the igneous rock that contains them.*

TABLE 1.3 FAULT TERMINOLOGY

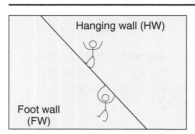

Hanging wall (HW) = the block of rock physically *above* the fault plane.

Foot wall (FW) = the block of rock physically *below* the fault plane.

Notice the little blue figures. Their hands are on the hanging wall and their feet are on the foot wall.

Normal Faults	Reverse Faults

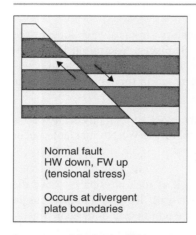

Normal fault
HW down, FW up
(tensional stress)

Occurs at divergent
plate boundaries

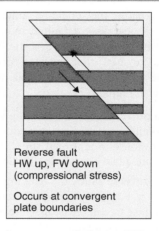

Reverse fault
HW up, FW down
(compressional stress)

Occurs at convergent
plate boundaries

In a normal fault, the HW moves down, and the FW moves up. Normal faults occur at divergent plate boundaries under tensional stress.

In a reverse fault, the HW moves up, and the FW moves down. Reverse faults occur at convergent plate boundaries under compressional stress.

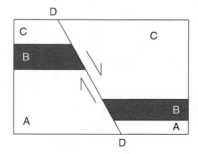

Unit A is the oldest, followed by B and C. Fault D is the youngest.

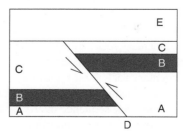

Unit A is the oldest, followed by B and C. Fault D is younger than C but older than unit E.

Figure 1.5 Normal faults, Thomasville Quarry, Thomasville, Pennsylvania.

Figure 1.6 Thrust fault. Interstate 59, Rising Fawn, Georgia.

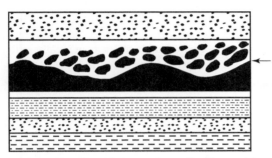

Figure 1.7 Erosion surface (indicated by arrow) overlain by a basal conglomerate.

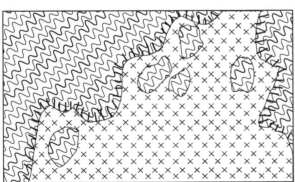

A. Intrusion of a plutonic igneous rock such as granite (indicated by x's) into a metamorphic rock. The intrusion contains xenoliths of the metamorphic rock.

Figure 1.8 Xenoliths.

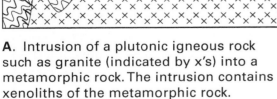

B. Xenolith of gneiss in the Stone Mountain Granite, Georgia. Pencil with scale in centimeters.

Principle of Fossil Succession

Where fossils are present in sedimentary rocks, the relative ages of the rocks can be determined from an examination of the fossils. This is because *fossils occur in a consistent vertical order in sedimentary rocks all over the world*. The **principle of fossil succession** is valid and does not depend on any preexisting ideas about evolution. Fossil succession was first recognized by William "Strata Bill" Smith in the late 1700s in England, more than 50 years before Charles Darwin published his theory of evolution. Geologists, however, interpret fossil succession to be the result of evolution, the natural appearance and disappearance of species through time.

SEDIMENTARY CONTACT AND UNCONFORMITIES

Types of Contact

There are two basic types of contacts between rock units: conformable and unconformable.

> **Conformable contacts** between beds of sedimentary rocks may be either abrupt or gradational. Most abrupt contacts are *bedding planes* resulting from sudden minor changes in depositional conditions. Gradational contacts represent more-gradual changes in depositional conditions. Conformable contacts generally indicate that no significant time gap or break in deposition has occurred.

> **Unconformable contacts** (or **unconformities**) are surfaces that represent a gap in the geologic record because of either *erosion* (see Principle of Components or Inclusions, earlier) or *nondeposition*. The time represented by this gap can vary widely, ranging up to millions or hundreds of millions of years (such as an erosional surface between Precambrian rocks and recent sediments). Unconformities are useful in relative dating because recognizing them allows us to distinguish between the older rocks below the unconformity and the younger rocks above the unconformity.

Unconformities may be recognized by a variety of types of criteria, including sedimentary features, fossils, and structural relationships (Box 1.2).

Box 1.2 Criteria for Recognizing Unconformities

Sedimentary Criteria

- Basal conglomerates (clasts may be from an underlying unit)
- Buried soil profiles
- Beds of phosphatized pebbles
- Glauconite (greensand)
- Manganese-rich beds

Paleontologic (Fossil) Criteria

- Abrupt changes in fossil assemblages (such as a change from marine to land-dwelling fossils, or the absence of certain fossil species that normally follow one another in sequential order)
- Bone or tooth conglomerates

Structural Criteria

- An irregular contact that cuts across bedding planes in the under-lying unit
- A difference in the angle of dip of the beds above and below the contact
- Truncation of dikes or faults along a sedimentary contact
- Truncation of igneous or metamorphic rock units along a sedimentary contact
- Above the unconformity, sedimentary units that are basically parallel with the unconformity surface

Types of Unconformities

There are four basic types of unconformities, distinguished by sedimentary criteria, such as erosional surfaces overlain by basal conglomerates, by paleontologic criteria such as abrupt changes in fossil assemblages, or by structural criteria such as irregular erosional surfaces that truncate igneous or metamorphic rocks.

Angular Unconformities

Angular unconformities are characterized by an erosional surface that truncates folded or dipping (tilted) strata. Overlying strata are deposited basically parallel to the erosion surface. The rocks above and below the unconformity are at an angle to one another. (Figure 1.9)

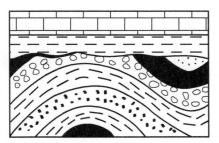
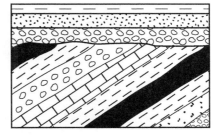

Figure 1.9 Angular unconformities.

Nonconformities

Nonconformities are characterized by an erosional surface that truncates igneous or metamorphic rocks. At a nonconformity, sedimentary rocks unconformably overlie igneous or metamorphic rocks (Figure 1.10).

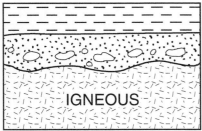
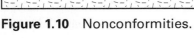
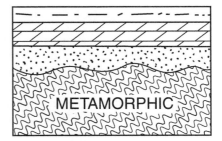

Figure 1.10 Nonconformities.

Disconformities

Disconformities are characterized by an irregular erosional surface that truncates flat-lying sedimentary rocks. The layers of sedimentary rocks above and below the unconformity are parallel to one another (Figure 1.11).

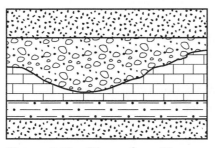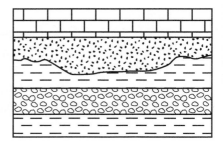

Figure 1.11   Disconformities.

Paraconformities

Paraconformities are characterized by a surface of nondeposition separating two parallel units of sedimentary rock. The surface is virtually indistinguishable from a sharp conformable contact, and there is no obvious evidence of erosion. An examination of the fossils, however, reveals that there is a considerable *time gap* between the parallel layers of sedimentary rock (Figure 1.12).

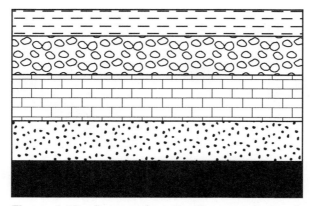

Figure 1.12   Paraconformity. You cannot see it without examining the fossils for a gap.

TYPES OF FOLDS

In the section on angular unconformities, we see that the rocks below the unconformity surface have been folded or tilted relative to the rocks above the unconformity. Folding and tilting of strata are caused by tectonic stresses within the Earth. For example, during mountain building, compressional stress can deform rocks to produce folds. Generally, a series of folds is produced much as a carpet might wrinkle when you push on one end. The up-folds and the down-folds are adjacent to one another, and grade into one another. In geology we give each a separate descriptive name: **anticline** (an up-fold, shaped like the letter A or an arch) and **syncline** (a U-shaped down-fold, shaped like your sink) (Table 1.4).

IGNEOUS CONTACTS

Igneous contacts can also be helpful in relative dating, particularly when considered along with the principle of cross-cutting relationships. If the contact is from an igneous rock intruding another rock type, a *contact metamorphic aureole* is present along the edge of the pluton (Figure 1.13). *Note:* In this lab manual, the contact metamorphic aureole is indicated by a series of short lines surrounding the igneous rock.

In some cases, an igneous rock such as basalt is interlayered between sedimentary strata. The question arises whether it is a *sill* or a *lava flow* (Table 1.5). A sill has a contact metamorphic aureole around all of its edges. A lava flow has evidence of contact metamorphism only along its lower side. The upper side may be marked by evidence of **subaerial** (exposed to the air) or **subaqueous** (under water) exposure, such as vesicles or small holes formed by gas bubbles escaping from the lava (Figure 1.14). These **vesicles** are called **amygdules** if they have been filled subsequently by minerals.

TABLE 1.4   ANTICLINES AND SYNCLINES

Anticline	Syncline

Anticline. (Geology student for scale.)

Syncline. Sideling Hill, Maryland. (Pedestrian bridge for scale.)

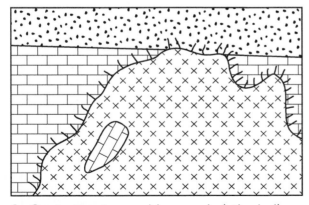

A. Contact metamorphic aureole (stippled) around a pluton (indicated by x's).

B. Triassic–Jurassic diabase dike (weathered yellow) surrounded by gray contact metamorphic aureole. The diabase dike has intruded into red siltstone. Deep River Basin, Borden Brick and Tile Quarry, Bilboa, Durham, North Carolina. (Field notebook and pencil for scale.)

Figure 1.13   Contact metamorphic aureoles.

TABLE 1.5   DISTINGUISHING A LAVA FLOW FROM A SILL

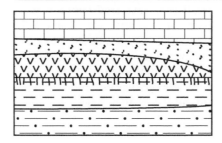 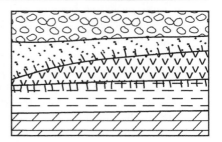

Lava flow (indicated by v's) with contact metamorphic aureole on the bottom (*stippled*).

Sill (indicated by v's) with contact metamorphic aureole along both top and bottom (*stippled*).

Pamela Gore

Figure 1.14   Vesicular basalt from a lava flow. The small holes, called *vesicles*, were formed by gas bubbles in the lava. (Scale in millimeters.)

Relative Dating Exercises

Examine the geologic cross sections and determine the relative ages of the rock bodies, lettered features such as intrusions, faults or surfaces of erosion, and other events such as tilting and folding. *Always start with the oldest rock and work toward the present.* List the letters in order, with the *oldest at the bottom.*

Example 1.1

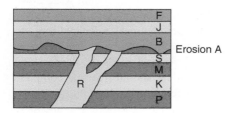

Think about each relative dating diagram as the side view of a layer cake. In general, the oldest units are on the bottom and the youngest units are on the top. There can be lots of complexities, such as folding, faulting, erosion, and intrusion by magma. You have to put these events into the order in which they occurred, starting with the oldest, and working toward the youngest. Figure out what cuts what. If a fault cuts a bed, then the bed is older than the fault.

In the figure above (Example 1.1), the sedimentary units are in the sequence P, K, M, and S; then something happened. A body of magma (Intrusion R) has intruded or cut through all of the previous layers, so it comes next in the sequence. The intrusion is eroded off at the top. (The previous layers are eroded off at the top too.) Thus the event *after* the intrusion is Erosion A. After Erosion A, beds B, J, and F were deposited.

The answer for this problem is: P, K, M, S, Intrusion R, Erosion A, B, J, F.

Example 1.2

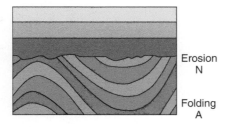

In the figure above (Example 1.2), there are no letters or numbers, but you could either label the diagram A, B, C, starting at the bottom or, even easier, describe the units in order.

You can see that units green, red, blue, green, red, blue, green, blue, green were deposited. Then Folding Event A occurred. Then Erosion N occurred. Then units brown, orange, and yellow were deposited. And that is the solution to the second problem.

INSTRUCTIONS

Solve the relative dating problems in the ten blocks below. Write the answers beside the blocks, with the oldest units at the bottom and the youngest units on top. Include all events (folding, faulting, etc.) in their proper sequence. The diagrams in the exercises are similar to those in the examples. If the block has units with letters on them, put the letters in order from oldest to youngest, as we did with the two examples. Add words for erosion events, folding events, intrusions, faults, tilting, etc. If a fault (or other event) is labeled with a letter, refer to it as "Fault A." If there are several faults and they are not labeled, you may say "the fault on the left" or the "fault on the right." If a layer does not have a label, determine the rock type from the lithologic symbol, or give its color. If the block is not labeled at all (Blocks 7 and 8), use the lithologic symbols to identify the rock types. Put the rock types and events (erosion, fault, folding, tilting, and intrusion) in order from oldest to youngest, as you did with the other diagrams.

In a classroom setting, this exercise works well if students work in pairs or small groups to solve the sequence of events in each block. Each student should write down the answers and be prepared to explain the sequence of events in each block. When everyone has completed the block diagrams, the instructor can ask a student to go to the front of the room to explain his or her logic in determining the sequence of events for a particular block. (If a visual presenter and projector are available, they will be useful to display the block under discussion. The student should give the complete sequence of events without interruption.) After the student finishes his or her explanation, the instructor should ask the class if anyone has anything different. Students in the audience should address any differences. (In some cases, there are two (or more) equally valid explanations if the sequence is ambiguous.) The instructor should give feedback at this point.

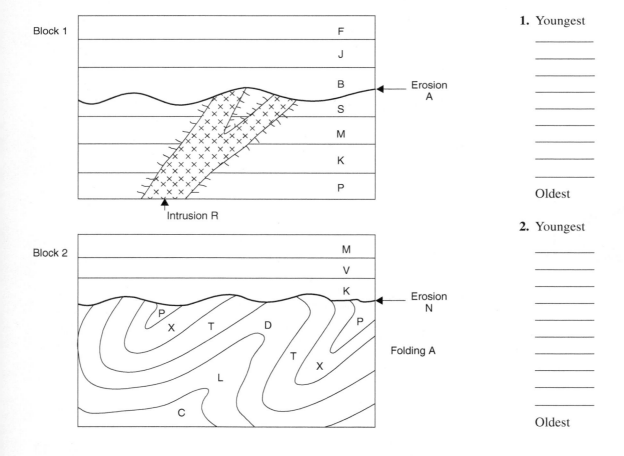

1. Youngest

_____
_____
_____
_____
_____
_____
_____
_____
_____

Oldest

2. Youngest

_____
_____
_____
_____
_____
_____
_____
_____
_____

Oldest

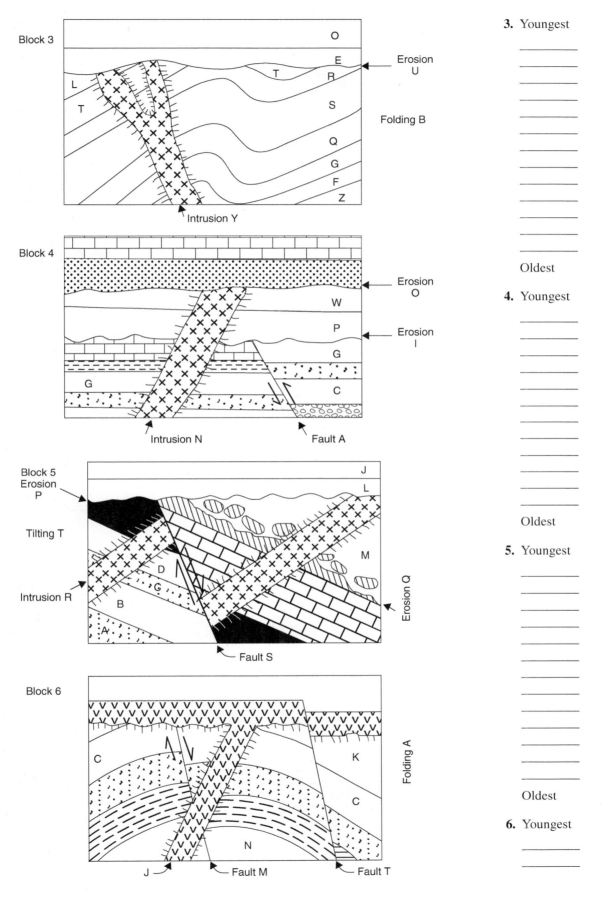

3. Youngest

_____

_____

_____

_____

_____

_____

_____

_____

_____

_____

_____

_____

Oldest

4. Youngest

_____

_____

_____

_____

_____

_____

_____

_____

_____

_____

_____

_____

Oldest

5. Youngest

_____

_____

_____

_____

_____

_____

_____

_____

_____

_____

_____

_____

Oldest

6. Youngest

_____

_____

Block 7

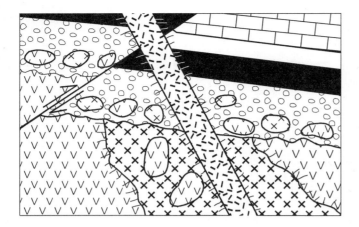

Block 8

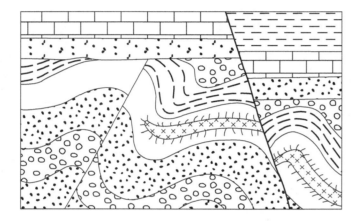

Block 9

Faulting U
Erosion R
Erosion E

_____
_____
_____
_____
_____
_____
_____

Oldest

7. Youngest

_____
_____
_____
_____
_____
_____
_____
_____
_____

Oldest

8. Youngest

_____
_____
_____
_____
_____
_____
_____
_____
_____
_____
_____
_____
_____

Oldest

9. Youngest

_____
_____
_____
_____
_____
_____
_____
_____
_____
_____
_____
_____

Oldest

Block 10

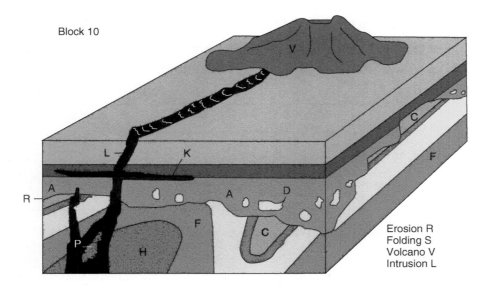

Erosion R
Folding S
Volcano V
Intrusion L

10. Youngest

————
————
————
————
————
————
————
————
————
————

Oldest

After determining the sequence of events in the ten diagrams above, answer the following questions.

1. What type of unconformity is represented by Erosion A in Block 1?

_____

2. What type of unconformity is represented by Erosion N in Block 2?

_____

3. What type of igneous intrusion is shown in Block 3?

_____

4. What type of fault is Fault A in Block 4?

_____

5. What type of fault is Fault S in Block 5?

_____

6. What type of fold is present in Block 6?

_____

7. In Block 6, is the igneous rock a pluton, a lava flow, or both?

_____

Explain your answer:

_____

8. In Block 7, are the inclusions with x's on them clasts, or are they xenoliths?

_____

9. In Block 7, are the inclusions with v's on them clasts, or are they xenoliths? Are all clasts with v's on them of that type, or are there other types as well?

_____

Explain your answer:

_____

10. What type of fault is present in Block 7?

_____

11. What types of faults are present in Block 8? Identify each.

Fault on left = _____

Fault on right = _____

12. Are there two faults in Block 9, or is there only one fault? _____

Explain your answer:

_____

13. What type of fault (or faults) is (or are) present in Block 9: normal or reverse?

_____

_____

14. In Block 9, what type of unconformity is found below unit L along erosion surface R?

_____

15. In Block 10, what type of unconformity is found below unit A along erosion surface R?

_____

16. What type of fold is the fold on the left in Block 10: an anticline or a syncline?

_____

17. What type of fold is the fold on the right in Block 10: an anticline or a syncline?

_____

18. Is the structure at P a clast or a xenolith? _____

19. Is the structure at D a clast or a xenolith? _____

20. Is the intrusion at L a dike or a sill? _____

21. Is the intrusion at K a dike or a sill? _____

Rocks and Minerals

Minerals are the building blocks of rocks, and rocks are the building blocks of Earth's crust. This lab introduces you to some of the major rock-forming minerals and to the three major groups of rocks: igneous, sedimentary, and metamorphic. Later labs in this course go into more detail about sediments and sedimentary rocks, because it is the sedimentary rocks that record the history of life on Earth. We can also learn to read the rock layers to learn many things about the conditions on Earth in the far geologic past, such as ancient climates, tectonic settings, and depositional environments.

BASIC DEFINITIONS

Minerals

Minerals are the building blocks of rocks. Each mineral has distinct physical and chemical properties that allow it to be identified. Minerals are naturally occurring inorganic solids with a definite chemical composition and an orderly internal crystal structure. These five characteristics that define a mineral are listed in Box 2.1. Examples of minerals are quartz, feldspar, calcite, and muscovite mica.

Box 2.1 Five Characteristics of Minerals
1. Naturally occurring
2. Inorganic
3. Solid
4. Definite chemical composition
5. Orderly internal crystal structure

Rocks

A **rock** is an aggregate of one or more minerals. Rocks are the building blocks of Earth's crust. Earth's continental crust is dominated by granite, and the oceanic crust is dominated by basalt. Both of these are **igneous rocks**, formed by the cooling of hot molten lava or magma. **Sedimentary rocks** cover about 75% of the surface of the continents, and they are formed from the weathered products of preexisting rocks. Shale and sandstone are the most abundant sedimentary rocks. Sedimentary rocks contain **fossils**, the preserved remains of organisms, which document the evolution of life through time. Rocks that have been changed by heat and/or pressure are called **metamorphic rocks**. Metamorphic rocks dominate the shield area of the continent and may be found in the cores of mountain belts. Schist and gneiss are common metamorphic rocks. The three basic categories of rocks are summarized in Table 2.1.

TABLE 2.1 CATEGORIES OF ROCKS

Category	Origin	Examples
Igneous	Crystallized from hot lava or magma	Granite, basalt
Sedimentary	Formed from sediment (e.g., gravel, sand, clay) that is derived from weathered rock	Sandstone, shale, limestone
Metamorphic	Rocks changed by heat and/or pressure	Gneiss, schist, slate, marble

PHYSICAL PROPERTIES OF MINERALS

Minerals have certain physical properties that help us identify them. The seven physical properties that you can readily determine in the laboratory are color, luster, hardness, cleavage, fracture, magnetism, and reaction to dilute hydrochloric acid.

Color

Color is the color of the mineral as it appears in reflected light to the naked eye.

Luster

Luster is the character of the light reflected from the mineral. A mineral may have a metallic luster (in other words, you would call it a metal), or a nonmetallic luster. The types of non-metallic luster are listed in Box 2.2.

Box 2.2 Types of Nonmetallic Luster

- Glassy or vitreous
- Dull
- Pearly
- Resinous (like amber)
- Waxy
- Adamantine (like a diamond)
- Silky

Hardness

Hardness is the resistance of a mineral to scratching. The hardness of minerals is measured on a scale of 1 to 10, the Mohs Hardness Scale (Table 2.2), with 1 the softest (talc) and 10 the hardest (diamond). In lab, we express the hardness of a mineral in comparison to common objects: fingernail (2.5), copper penny (2.9), wire or iron nail (4.5), glass (5.5), and ceramic streak plate (6.5). A fingernail can scratch gypsum. Potassium feldspar, quartz, and minerals of greater hardness can scratch glass.

TABLE 2.2   MOHS HARDNESS SCALE

Mohs Hardness Scale	Mineral	Hardness Comparator
1	Talc	
2	Gypsum	
		Fingernail (2.5)
		Copper penny (2.9)
3	Calcite	
4	Fluorite	
		Wire or iron nail (4.5)
5	Apatite	
		Glass (5.5)
6	Orthoclase feldspar (potassium feldspar)	
		Ceramic streak plate (6.5)
7	Quartz	
8	Topaz	
9	Corundum	
10	Diamond	

Cleavage

Cleavage is the tendency of a mineral to break along flat surfaces related to planes of weakness in its crystal structure. Minerals can be identified by the number of cleavage planes they exhibit and the angles between the cleavage planes. For example, some minerals tend to cleave or break into flat sheets (the micas: muscovite and biotite), others break into cubes (halite) or into rhombs (calcite and dolomite). Other minerals have different types of cleavage. For this lab, it is necessary to recognize whether or not minerals have cleavage and to tell if the cleavage is one of these types.

Fracture

Fracture is irregular breakage not related to planes of weakness in the mineral. Some minerals, such as quartz and olivine, do not have cleavage. Instead, they have a special type of breakage called conchoidal fracture. **Conchoidal fracture** produces curved breakage surfaces, such as would be seen on arrowheads or chipped glass.

Magnetism

Magnetism is a property of a few minerals. Magnetic minerals are attracted to a magnet, or they act as a natural magnet, attracting small steel objects such as paper clips. The only magnetic mineral that you might see in lab is **magnetite**.

Reaction to Acid

Reaction to acid is a physical property of some minerals. The carbonate minerals react with dilute **hydrochloric acid** (HCl) by effervescing or fizzing, producing bubbles of carbon dioxide gas (the same type of gas bubbles that are found in carbonated beverages). **Calcite** fizzes readily in hydrochloric acid. **Dolomite** fizzes if it is first scratched and powdered. You may use a nail or steel needle probe to scratch the specimen to try this test. (See the section near the beginning of this lab manual entitled "Guidelines for Using Acid to Identify Minerals.")

CHEMICAL COMPOSITION OF MINERALS

Minerals can be divided into groups based on their chemical composition. For this lab, we will divide the minerals into the **silicate** minerals (those containing the elements silicon and oxygen; Table 2.3) and the **nonsilicate** minerals (Table 2.4. There is a wide variety of nonsilicate minerals, including **carbonate minerals** (those containing the carbonate ion, CO_3^{2-} in their chemical formula) along with halides, sulfates, sulfides, and oxides (which we will group into "other" in this lab). Silicates are the dominant minerals in igneous rocks, and they dominate many types of sedimentary and metamorphic rocks. In the metamorphic rock section of this lab you will see photos of some silicate minerals that form during metamorphism (Tables 2.5 and 2.6). Use Tables 2.3–2.6 to identify minerals in the lab.

TABLE 2.3 SILICATE MINERALS					
Mineral	**Color**	**Hardness***	**Luster**	**Cleavage**	**Reaction to HCl**
Quartz	Colorless, gray, white, pink, brown, etc.	H > glass H = 7	Nonmetallic glassy	Conchoidal fracture No cleavage	—
Feldspar Group (potassium feldspar, and plagioclase feldspar)	White, pink, gray, green	H > glass H = 6	Nonmetallic glassy	Has cleavage Two directions of cleavage at 90 degrees	—
Olivine	Olive green	H > glass H = 6.5–7	Nonmetallic glassy	Conchoidal fracture No cleavage	—
Muscovite mica	Colorless, tan, silver	H < fingernail H = 2–2.5	Nonmetallic glassy to pearly	Splits into flat sheets	—
Biotite mica	Brown to black	H > fingernail H < penny H = 2.5–3	Nonmetallic glassy	Splits into flat sheets	—

Mineral	Color	Hardness*	Luster	Cleavage	Reaction to HCl
Pyroxene	Green to black	H < nail H = 5–7	Nonmetallic glassy	Has cleavage Two directions of cleavage at 90 degrees	—
Amphibole	Green to black	H < nail H = 5–6	Nonmetallic glassy to silky	Has cleavage Two directions of cleavage not at 90 degrees	—
Garnet	Dark red	H > glass H = 6.5–7.5	Nonmetallic glassy to resinous	No cleavage	—
Clay minerals (e.g., kaolinite)	Variable, such as white, orange	H < fingernail H = 2	Nonmetallic dull	Cleavage not visible to naked eye	—

*H> indicates "hardness greater than" (i.e., harder than); H< indicates "hardness less than" (i.e., softer than).

TABLE 2.4 NONSILICATE MINERALS					
Mineral	Color	Hardness	Luster	Cleavage	Reaction to HCl
Carbonate Minerals					
Calcite	White, tan, gray, black, etc.	H > fingernail H < penny H = 3	Nonmetallic	Has cleavage Breaks into rhombs	Fizzes readily
Dolomite	White, tan, gray, black, etc.	H > penny H < nail H = 3.5–4	Nonmetallic	Has cleavage Breaks into rhombs	Fizzes if scratched and powdered
Other Nonsilicate Minerals					
Gypsum	Usually white	H < fingernail H = 2	Nonmetallic	Has cleavage Can break into fibers or flat sheets	—
Halite	Colorless to white or gray	H > fingernail H < penny H = 2.5	Nonmetallic	Has cleavage Breaks into cubes	—
Magnetite	Black to gray	H > glass H = 6	Submetallic to metallic	No cleavage	—
Pyrite	Brassy or gold	H > glass H = 6–6.5	Metallic	No cleavage	—
Hematite (iron oxide)	Reddish brown to silver to gray	Variable H = 5.5–6.5	Nonmetallic to metallic	No cleavage	—

TABLE 2.5   SILICATE MINERALS

Quartz is commonly white, pink, brown, gray, or colorless.

Feldspar has two directions of cleavage at 90 degrees.

Muscovite mica splits into flat, transparent sheets.

Biotite mica is black or brown and splits into flat sheets.

Pyroxene has two directions of cleavage at 90 degrees. (Scale in centimeters.)

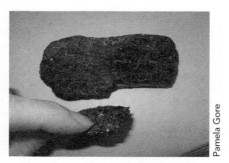

Amphibole has two directions of cleavage not at 90 degrees.

TABLE 2.6   NONSILICATE MINERALS

Calcite has rhombohedral cleavage and fizzes in hydrochloric acid.

Dolomite has rhombohedral cleavage but only fizzes if scratched or powdered. (Scale in millimeters.)

Gypsum has rhombohedral cleavage, but it is softer than a fingernail. (Scale in centimeters.)

Halite has cubic cleavage. (Scale in centimeters.)

Hematite has a reddish brown streak. Its luster can be dull or metallic.

Pyrite has a gold color, metallic luster, and cubic crystals but no cleavage. (Scale in centimeters.)

Muscovite and biotite are two types of **mica**. Micas are minerals that can be peeled apart in very thin layers to form flat sheets. This property of splitting into sheets is called cleavage. Micas have one direction of cleavage. **Muscovite mica** is silver, and **biotite mica** is black to brown.

There are two types of feldspar, as mentioned in Table 2.3. These are potassium feldspar and plagioclase feldspar. Both have two directions of cleavage (or shiny flat surfaces) at 90 degrees to each other. **Potassium feldspar** (sometimes called microcline or orthoclase feldspar) is typically pink, tan, white, or green with narrow white stripes (see the photo in Table 2.5). **Plagioclase feldspar** (actually a group of feldspar minerals containing sodium and/or calcium) is typically white, gray or black and can have flashes of iridescent blue or green when turned at various angles. On close inspection with a hand lens or microscope, you can see tiny parallel grooves on some flat cleavage planes. You will not need to distinguish between these two types of feldspar in this lab.

Amphibole and pyroxene are black to dark green or dark brown minerals that are typically somewhat elongated and are very similar in appearance. They both have two directions of cleavage; however, the angle between the cleavage planes differs in each. In cross section, **amphibole** has a shape similar to a diamond on a playing card, with sides that meet at 60-degree and 120-degree angles. In contrast, **pyroxene** has a square cross section with sides that meet at 90-degree angles. For simplicity, in this lab, they can be lumped together under the general term **pyrobole** (which comes from *pyro*xene and amphi*bole*) because it is often difficult for introductory students to distinguish between them.

IGNEOUS ROCKS

Igneous rocks are "fire formed" (Latin ignis, "fire"). They crystallized from hot, molten lava or magma as it cooled. **Magma** is hot molten rock beneath the surface of the Earth. **Lava** is hot, molten rock that has flowed out onto the surface of the Earth. Magma can cool within the Earth's crust to form igneous rocks. Lava cools more quickly because it is on the Earth's surface, where temperatures are much cooler than they are at depth.

Most igneous rocks consist of a mosaic of intergrown crystals of silicate minerals that formed as the hot molten magma or lava cooled. These minerals typically include feldspar, quartz, amphibole, pyroxene, olivine, biotite, and muscovite.

Cooling rates influence the texture of the igneous rock (Table 2.7).

Quick cooling yields fine grains, Slow cooling yields coarse grains. Some igneous rocks, such as obsidian and pumice, cooled virtually instantaneously to form a natural glass, and because of this, they lack minerals.

Composition of Igneous Rocks

Igneous rocks can be classified into four groups based on their chemical compositions: sialic, intermediate, mafic, and ultramafic (Table 2.8). In general, these groups can be distinguished on the basis of color. **Sialic** igneous rocks are light in color and are commonly white, light gray, or pink. **Intermediate** igneous rocks are gray or a mixture of half black minerals and half white minerals. **Mafic** igneous rocks are black to dark gray, but they can weather to red, brown, or orange. **Ultramafic** igneous rocks may be bright olive green if they are dominated by the mineral **olivine**.

TABLE 2.7   TEXTURES OF IGNEOUS ROCK

Glassy – Glassy texture results from instantaneous cooling.
Example: Obsidian.

Obsidian (volcanic glass). (Scale in centimeters.)

Vesicular – Rocks with a vesicular texture are very porous. They contain tiny holes called vesicles that formed from gas bubbles in lava or magma.
Examples: Pumice, vesicular basalt, scoria.

Pumice is spongelike and floats on water. It is used as an abrasive in products like pumice stone and some soap. (Scale in centimeters.)

Aphanitic – Fine grain size (<1 mm). Aphanitic texture is the result of quick cooling.
Examples: Basalt, rhyolite, andesite.

Rhyolite. (Scale in centimeters.)

(Continued)

TABLE 2.7 *(Continued)*

Phaneritic – Coarse grain size, visible grains (1–10 mm). Phaneritic texture is the result of slow cooling.
Examples: Granite, diorite, gabbro.

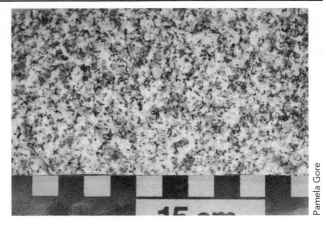

Granite. (Scale in centimeters.)

Pegmatitic – Very coarse grain size. Crystals may be several inches across.
Example: Granite pegmatite.

Granite pegmatite vein cutting through granite. (Scale in centimeters.)

Porphyritic – A mixture of grain sizes. Larger crystals (called phenocrysts) are present in a finer grained phaneritic or aphanitic igneous rock.
Examples: Porphyritic granite, porphyritic andesite, porphyritic basalt.

Porphyritic granite with large pink phenocrysts of potassium feldspar. (Scale in millimeters.)

TABLE 2.8 COMPOSITION OF IGNEOUS ROCK

Sialic (or granitic or felsic)

- Dominated by silicon and aluminum (SiAl)
- Usually light in color (except obsidian)
- Characteristic of continental crust
- Forms a stiff (viscous) lava or magma
- Minerals commonly present include:
  - ◊ Potassium feldspar (pink or white)
  - ◊ Plagioclase feldspar (generally white)
  - ◊ Quartz (generally gray or colorless)
  - ◊ Biotite mica
  - ◊ Amphibole (may be present)
  - ◊ Muscovite (may be present)

- Rock types include:
  - ◊ Rhyolite (aphanitic texture)
  - ◊ Granite (phaneritic texture)
  - ◊ Obsidian (glassy texture)
  - ◊ Pumice (vesicular texture)
  - ◊ Granite pegmatite (pegmatitic texture)

Granite. Note the pink potassium feldspar, white plagioclase feldspar, gray quartz, and black biotite mica. (Scale in centimeters.)

Intermediate (or andesitic)

- Intermediate in composition between sialic and mafic
- Minerals commonly present include:
  - ◊ Plagioclase feldspar
  - ◊ Amphibole (black)
  - ◊ Pyroxene (black)
  - ◊ Biotite (black)
  - ◊ Quartz

- Rock types include:
  - ◊ Andesite (aphanitic texture)
  - ◊ Porphyritic andesite (porphyritic aphanitic texture)
  - ◊ Diorite (phaneritic texture)

Note: Andesite can have a few larger black crystals (phenocrysts) in the fine-grained, gray rock (porphyritic texture). The larger crystals are called phenocrysts, and in andesite they are generally the mineral amphibole.

Porphyritic andesite with phenocrysts of black amphibole. (Scale in millimeters.)

(Continued)

TABLE 2.8 *(Continued)*

Diorite has a phaneritic texture with approximately half black minerals and half white minerals. (Scale in millimeters.)

Mafic (or basaltic)

- Contains abundant ferromagnesian minerals (iron and magnesium silicates)
- Usually dark in color (dark gray to black)
- Characteristic of Earth's oceanic crust and Hawaiian volcanoes
- Basalt forms runny (low-viscosity) lava
- Basalt is also found on the Moon, Mars, and Venus
- Minerals commonly present include:
  - ◊ Amphibole
  - ◊ Pyroxene
  - ◊ Plagioclase feldspar
  - ◊ Olivine
- Rock types include:
  - ◊ Basalt (aphanitic texture)
  - ◊ Vesicular basalt
  - ◊ Porphyritic basalt
  - ◊ Diabase
  - ◊ Gabbro (phaneritic texture)

Basalt has an aphanitic texture and a dark gray to black color. (Scale in centimeters.)

Gabbro has a phaneritic texture and a dark gray to black or brown color. (Scale in millimeters.)

Ultramafic

- Dominated almost entirely by ferromagnesian minerals (iron and magnesium silicates)
- Rarely observed on Earth's surface
- Major constituent of Earth's mantle
- May be found as xenoliths or inclusions in basaltic rocks
- Minerals commonly present include olivine (has an olive green color)
- Can have minor amounts of:
    ◊ Pyroxene
    ◊ Plagioclase feldspar
- Rock types include: peridotite (phaneritic texture)

Note: Peridotite is dominated by the mineral olivine. Olivine is olive green. The birthstone peridot is olivine and gives its name to peridotite.

Peridotite is an ultramafic rock dominated by the green mineral olivine. (Scale in millimeters.)

Classification of Igneous Rock

Igneous rocks are classified or named based on their **texture** and their **composition**. Use Table 2.9 to help you identify the igneous rocks. Determine the texture and the general composition of the rock specimen (based on color), and then read its name from Table 2.9. Other igneous rocks, that you might see in lab but are not listed in the table, include obsidian and pumice.

TABLE 2.9 CLASSIFICATION OF IGNEOUS ROCK				
Texture	**Composition**			
	Sialic	**Intermediate**	**Mafic**	**Ultramafic**
Aphanitic (fine grained)	Rhyolite	Andesite	Basalt	—
Phaneritic (coarse grained)	Granite	Diorite	Gabbro	Peridotite

SEDIMENTARY ROCKS

Sedimentary rocks are made from sediment. **Sediment** is loose particulate material (such as gravel, sand, or clay) derived from the weathering of rocks. Sediment is transported by water, wind, or ice and deposited in sedimentary environments such as beaches, rivers, deserts, lakes, and the sea floor.

There are also chemical and biochemical sedimentary rocks (such as evaporites and carbonate rocks) made from the chemical weathering products of rocks. Organic sedimentary rocks (coal) are composed of the carbon-rich remains of plants. Sedimentary rocks commonly contain **fossils**, which are the remains of once-living organisms.

Sediment becomes sedimentary rock through **lithification**, which involves the processes of compaction, cementation, or recrystallization (of carbonate sediment). The major types of sedimentary rocks are terrigenous, chemical or biochemical and organic sedimentary rocks.

Terrigenous Sedimentary Rocks

Terrigenous (also called clastic or detrital) sedimentary rocks are derived from the weathering of preexisting rocks (Table 2.10). They have a clastic (broken or fragmental) texture consisting of clasts, matrix, and cement. Terrigenous sedimentary rocks are classified according to their texture or the grain size of the sediment as gravel, sand, silt, or clay.

TABLE 2.10 TERRIGENOUS ROCKS

Components of a terrigenous or clastic sedimentary rock are:

- Clasts: larger pieces, such as grains of sand or gravel
- Matrix: mud or fine-grained sediment surrounding the clasts
- Cement: the glue that holds it all together, such as:
  - ◊ Calcite: fizzes in hydrochloric acid
  - ◊ Iron oxide: reddish brown
  - ◊ Silica: does not fizz and is not reddish brown

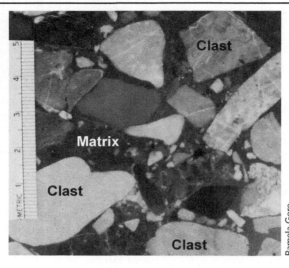

Clasts and matrix (labeled) and iron oxide cement (reddish brown). (Scale in millimeters.)

Gravel has a grain size greater than 2 mm.

Conglomerate has rounded clasts. (Penny for scale.)

Breccia has angular clasts. (Penny for scale.)

Sand has a grain size of 1/16 to 2 mm. Sandstone is commonly made of quartz and feldspar, but it can contain grains of nearly any mineral composition or grains of fine-grained rocks (e.g., basalt or shale).

Quartz sandstone (also called quartz arenite) is dominated by quartz grains. (Scale in centimeters.)

Additional types of sandstone. Sandstone with abondant fields per grains is called arkose. Sandstone with abondant rock fragment grains is called in this sandstone litharenite, or graywacke.

Arkose is dominated by sand-sized feldspar grains. (Scale in millimeters.)

Lithic sandstone (also called litharenite or graywacke) is dominated by sand-sized rock fragment grains. (Scale in centimeters.)

Silt is intermediate in size between sand and clay, with a grain size ranging from 1/256 to 1/16 mm. Siltstone feels gritty to the fingernails.

Siltstone is dominated by silt-sized grains, which are smaller than sand grains. (Scale in centimeters.)

Clay has a grain size less than 1/256 mm, and feels smooth to the fingernails. Sedimentary rocks with clay-sized grains may be shale, which is fissile, breaking into flat layers, or claystone, which is massive, breaking into irregular blocks.

Mud is technically a mixture of silt and clay. It forms a rock called mudstone if it is massive or mudshale if it is fissile (splits into flat layers).

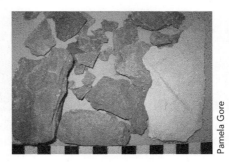

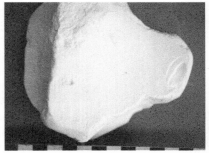

Shale is fissile. It breaks into flat layers. (Scale in centimeters.)

Claystone is massive. It breaks into irregular blocks. White claystone is called kaolin. (Scale in centimeters.)

Chemical and Biochemical Sedimentary Rocks

The group of chemical and biochemical sedimentary rocks includes the **evaporites**, the **carbonates** (limestones and dolostone), the **siliceous rocks**, and the **sedimentary ironstones**.

These rocks form within the depositional basin from chemical components dissolved in the seawater. These chemicals may be removed from seawater and made into rocks by chemical processes or with the assistance of biological processes (such as shell growth). In some cases it is difficult to sort the two out (in carbonates or some siliceous rocks, for example), so they are grouped together as chemical/biochemical.

Evaporites

Evaporites form from the evaporation of water that contains dissolved salts or minerals (such as seawater or salt lakes). The salts are derived from the chemical weathering of rocks. Evaporites include **rock salt**, **rock gypsum**, and **travertine**.

Rock salt is composed of the mineral halite (NaCl, sodium chloride) (Figure 2.1).

Rock gypsum is composed of the mineral gypsum ($CaSO_4 \cdot 2H_2O$). You can scratch it with your fingernail (Figure 2.2).

Travertine is an evaporite because it forms from the evaporation of water, and it is a carbonate rock because of its chemical composition, calcium carbonate ($CaCO_3$). Travertine forms in caves as stalactites and stalagmites, and it forms around hot springs from the evaporation of groundwater rich in dissolved calcium carbonate (Figure 2.3).

Figure 2.1 Rock salt. (Scale in millimeters.)

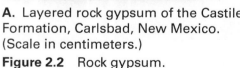

A. Layered rock gypsum of the Castile Formation, Carlsbad, New Mexico. (Scale in centimeters.)

Figure 2.2 Rock gypsum.

B. Gypsum crystals. Marion Lake, Australia. (Scale is 15 centimeters.)

Figure 2.3 Travertine stalagmites and stalactites. Cave of the Winds, Manitou Springs, Colorado.

Carbonate Rocks

The carbonate sedimentary rocks are formed through both chemical and biochemical processes. Limestone and dolostone are carbonate sedimentary rocks. Limestone is a sedimentary rock dominated by the mineral calcite. There are many types of limestone based on their textures, including fossiliferous limestone, chalk, and coquina (Figure 2.4). (You will learn more about the types of limestone in a later lab.) Dolostone is a sedimentary rock dominated by the mineral dolomite.

Two minerals are dominant in carbonate rocks. Calcite ($CaCO_3$) fizzes readily in hydrochloric acid. Dolomite ($CaMg(CO_3)_2$) fizzes in hydrochloric acid only if scratched or powdered.

Pamela Gore

A. Fossiliferous limestone with brachiopod fossils. (Specimen is about 12 cm wide.)

15 cm

Pamela Gore

B. Coquina with broken and rounded pieces of mollusc (clam) shells. This is a carbonate rock with a clastic texture. (Scale in centimeters.)

Figure 2.4 Types of limestone.

Siliceous Rocks

The siliceous rocks, such as diatomite and chert, are dominated by silica (SiO_2) (Figure 2.5). They commonly form from silica-secreting organisms such as diatoms, radiolarians, or some types of sponges. (You will learn about these organisms in later labs).

Diatomite is made of the siliceous skeletons of fossilized microscopic planktonic organisms called diatoms. It looks like chalk, but it does not fizz in acid. It can also resemble kaolinite, but diatomite is more porous, much lower in density and may float on water. Diatomite is also referred to as diatomaceous earth.

Chert is massive, hard microcrystalline quartz. Chert does not fizz in acid, and may be dark or light in color. Chert commonly forms nodules or beds, replacing limestone.

A. Diatomite is white, does not fizz in acid, and has a very low density.
(Scale in millimeters.)

B. Chert is massive and hard, will scratch glass, and does not fizz in acid.
(Scale in millimeters.)

Figure 2.5 Siliceous rocks.

Sedimentary Ironstones

The sedimentary ironstones are dominated by iron-bearing minerals, such as hematite or limonite. Common examples of sedimentary ironstones include oolitic ironstone, iron oxide concretions, and Precambrian banded iron formations (Figure 2.6). Some of these rocks can be used as iron ore for steel making.

A. Oolitic hematite or oolitic ironstone. This specimen is of Silurian age from Red Mountain in Birmingham, Alabama. (Scale in centimeters.)

B. Iron oxide concretions in sandstone, Lumpkin, Georgia. These concretions are common in some areas of Cretaceous and Cenozoic sediments in the Coastal Plain. (Scale in centimeters.)

C. Banded iron formation of Precambrian age. These rocks have thin alternating red and black layers. (Scale in centimeters.)

Figure 2.6 Sedimentary ironstones.

Organic Sedimentary Rocks (Coals)

Coal is an organic sedimentary rock consisting primarily of carbon-rich organic matter from plant fragments. Coal is important because it is a major energy source, used to generate electricity. There are several types (or ranks) of coal including peat, lignite, bituminous coal and anthracite coal (Figure 2.7).

Peat is brown and has visible pieces of plant material that can easily be broken apart.

Lignite is soft, dull, brown or gray to black coal. It is brittle and crumbles easily. Black powdery soot rubs off on your fingers. Carbonized plant fragments might be visible.

Bituminous coal is hard black coal that is shiny, brittle, slightly sooty, and may have visible layers.

Anthracite coal (technically a metamorphic rock) is hard, black, and very shiny with a silvery submetallic luster. It can break with conchoidal fracture, and is not sooty.

With increasing depth of burial, temperature, and pressure, plant fragments are altered to the various types of coal:

Plant fragments → Peat → Lignite → Bituminous coal → Anthracite coal

A. Peat is a type of coal consisting of a mass of compacted brown plant fragments. (Scale in centimeters.)

B. Lignite is a type of coal that is soft, crumbles easily, and has powdery black soot.

C. Bituminous coal is shiny and can have visible layers. (Scale in centimeters.)

D. Anthracite coal (metamorphosed coal) is very shiny with a silvery sub-metallic luster, and conchoidal fracture. (Scale in centimeters.)

Figure 2.7   Types of coal.

METAMORPHIC ROCKS

Metamorphism means "changed form." Metamorphism changes rocks as a result of heat and/or pressure and the action of chemical fluids. High pressures on a rock can occur with deep burial, the pressure between colliding continents, or along fault zones. High temperatures can result from the heat of magma intruding into other types of rocks (contact metamorphism) or from deep burial. Metamorphism can affect sedimentary rocks, igneous rocks, or even other metamorphic rocks (Table 2.11).

TABLE 2.11 COMMON METAMORPHIC ROCKS AND THEIR PARENT ROCKS	
Metamorphic Rock	**Parent Rock**
Marble	Limestone
Dolomitic marble	Dolostone
Quartzite	Quartz sandstone
Slate	Shale
Phyllite	
Schist	
Gneiss	Shale or granite
Metabasalt	Basalt
Greenstone	
Amphibolite	Basalt, gabbro, or volcanic sediment
Anthracite coal	Peat, lignite, or bituminous coal

Metamorphism changes the texture and mineralogy of rocks. Pressure on the rock causes **compaction** (which moves the grains closer together, reducing porosity and making the rock more dense) and **recrystallization** (which involves of the growth of new crystals, with no change in the overall chemistry of the rock). New crystals grow from the minerals already present. Directed pressure causes a preferred orientation of platy or sheetlike minerals, such as muscovite and biotite mica, perpendicular to the direction of pressure. This preferred orientation of minerals is called **foliation**.

Metamorphic Minerals

A number of metamorphic minerals form during metamorphism, and they are found almost exclusively in metamorphic rocks (Figure 2.8). Muscovite and biotite, however, are commonly present in igneous rock. Muscovite can also be present in sand and sedimentary rock.

Garnet: Dark red dodecahedrons (12-sided crystals).

Micas: Muscovite is silvery, biotite is dark brown.

Kyanite: Sky-blue elongated minerals with differential hardness. They can be scratched lengthwise with a knife or nail, but not sideways.

Staurolite: Brown lozenge-shaped minerals, commonly twinned or intergrown to form "fairy crosses."

Tourmaline: Commonly black. Forms elongated crystals with a rounded triangular cross section.

A. Dark red garnet crystals in mica schist. On display at Tellus Science Museum, Cartersville, Georgia.

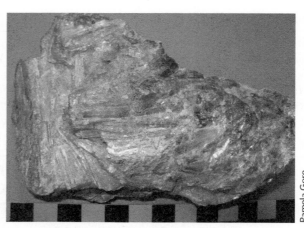

B. Blue kyanite crystals in quartzite. (Scale in centimeters.)

C. Brown staurolite crystals in mica schist. On display at Tellus Science Museum, Cartersville, Georgia.

D. Staurolite crystals, about 2 cm long, integrown to form "fairy crosses". On display at Tellus Science Museum, Cartersville, Georgia.

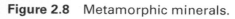

E. Black tourmaline crystals in quartzite. (Scale in millimeters.)

F. Talc.

Figure 2.8 Metamorphic minerals.

Talc: White or pale green and soft. Can be scratched with a fingernail. Commonly used in talcum powder.

Chlorite: Dark bluish green and soft. Contains magnesium or iron and is found in metamorphosed mafic igneous rocks.

Graphite: Metamorphosed carbon. Commonly used as pencil leads.

Metamorphic Textures

The most common metamorphic textures are foliated and nonfoliated.

Foliated metamorphic rocks are those that have a parallel alignment of sheetlike minerals, such as muscovite, biotite, or chlorite (Table 2.12). Examples include slate, phyllite, schist, and gneiss. As the clay-rich sedimentary rock, shale, is subjected to increasing grade of metamorphism (increasing temperatures and pressures), it undergoes successive changes in texture associated with recrystallization of the clays to mica and an increase in the size of the mica flakes. Slate, phyllite, schist, and gneiss form from the metamorphism of shale, with increasing temperatures and pressures (Table 2.13).

TABLE 2.12   FOLIATED METAMORPHIC ROCKS

Slate is a very fine-grained metamorphic rock. It resembles its parent rock, shale. It has slaty cleavage that may be at an angle to the original bedding. Relict bedding (bedding left over from the original sedimentary rock) may be visible on cleavage planes (shown here). Slate is commonly dark gray. It rings when you strike it (unlike shale, which makes a dull sound).

Slate with relict bedding (vertical). (Scale in centimeters.)

Phyllite (pronounced "fill-ite") is a fine-grained metamorphic rock with a frosted sheen, resembling frosted eye shadow or metallic paint on a car. This is no coincidence. Cosmetics and metallic paint commonly contain ground-up muscovite similar in size to that occurring naturally in phyllite.

Phyllite. (Scale in centimeters.)

(Continued)

TABLE 2.12 *(Continued)*

Schist is a metamorphic rock containing conspicuous muscovite and/or biotite mica flakes several millimeters across. Several types of schist may be recognized, based on the minerals present:

- Mica schist (shown here)
- Garnet schist
- Chlorite schist
- Kyanite schist
- Talc schist

Schist. (Scale in centimeters.)

Gneiss (pronounced "nice") is a banded or striped rock with alternating layers of dark and light minerals. The dark layers commonly contain biotite, and the light layers commonly contain quartz and feldspar.

Gneiss. (Scale in centimeters.)

TABLE 2.13   METAMORPHIC ROCKS THAT FORM FROM SHALE WITH INCREASING TEMPERATURE AND PRESSURE

Sedimentary rock	Metamorphic rocks Increasing temperature and pressure

Shale →	Slate → Phyllite → Schist → Gneiss

Nonfoliated metamorphic rocks have a granular, crystalline texture (Table 2.14). They are composed of equidimensional grains, such as quartz or calcite. There is no preferred orientation. The grains form a mosaic. Examples include quartzite and marble.

Quartzite forms from the metamorphism of quartz sandstone. Marble forms from the metamorphism of limestone. Metabasalt, greenstone, and amphibolite form from the metamorphism of basalt. These rocks can have a weak foliation if they contain micas or chlorite. If they contain abundant amphibole crystals that are aligned parallel with one another (such as in some amphibolites), and they can have a lineated texture. Anthracite coal is a nonfoliated metamorphosed coal.

Quartzites and marbles are not necessarily "pure" quartz or calcite. Some contain layers of impurities that were originally clay minerals, silt, or iron oxides interlayered or mixed with the original quartz sand or lime mud (Figure 2.9). These impurities metamorphose to form layers of micas or other minerals, which can give marble a banded, gneissic appearance or that can give a weak foliation to some quartzites (Figure 2.10).

Figure 2.9 Impurities such as clay in lime mud metamorphose to give marble a layered, gneissic appearance or a weak foliation. Three varieties of marble are present in the floor tiles at the Carlos Museum, Emory University, Atlanta, Georgia.

Figure 2.10 Clays mixed with sand can metamorphose to form layers of silver-colored muscovite in quartzite, giving it a weak foliation. Metamorphic minerals, such as the black tourmaline seen here, can also form in quartzite. (Setters Quartzite, Maryland. Scale in centimeters.)

TABLE 2.14   NONFOLIATED AND WEAKLY FOLIATED METAMORPHIC ROCKS

Marble fizzes in acid because its dominant mineral is calcite (or dolomite in dolomitic marble). It can be white, gray, or pink. The parent rock is limestone (or dolostone).

Marble. (Scale in centimeters.)

Quartzite consists of interlocking grains of quartz. The rock fractures through the grains (rather than between the grains, as it does in sandstone). The parent rock is quartz sandstone.

Quartzite. (Scale in centimeters.)

Greenstone is dark green rock with very fine texture. It can contain chlorite and may be massive to weakly foliated. The parent rock is basalt.

Greenstone. (Scale in centimeters.)

Metabasalt is massive black, greenish, or bluish-gray rock, which sometimes has vesicles that have been filled by minerals. Filled vesicles are called amygdules. The green mineral in the amygdule shown here is epidote. The parent rock is basalt.

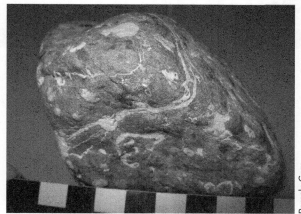

Metabasalt with amygdules. Syria, Virginia. (Scale in centimeters.)

Amphibolite is black rock dominated by amphibole with some plagioclase feldspar. It may be weakly foliated or lineated if the amphibole grains are parallel. The parent rock is basalt, gabbro, or volcanic sediment.

Amphibolite. (Scale in millimeters.)

Summary

Here is a list of the rocks and minerals mentioned in this lab.

I. MINERALS

SILICATE MINERALS

Rock-Forming Minerals

1. Feldspar (potassium feldspar and plagioclase feldspar)
2. Quartz
3. Muscovite mica
4. Biotite mica
5. Amphibole
6. Pyroxene
7. Olivine
8. Clay minerals (such as kaolinite)

Metamorphic Silicate Minerals

1. Garnet
2. Staurolite
3. Kyanite
4. Chlorite
5. Talc
6. Tourmaline

NONSILICATE MINERALS

1. Calcite
2. Dolomite
3. Halite
4. Gypsum
5. Pyrite
6. Magnetite
7. Hematite
8. Graphite

II. ROCKS

IGNEOUS ROCKS

Sialic

1. Granite
2. Rhyolite
3. Pumice
4. Obsidian

Intermediate

1. Diorite
2. Andesite

Mafic

1. Gabbro
2. Basalt
3. Vesicular basalt

Ultramafic

4. Peridotite

SEDIMENTARY ROCKS

Terrigenous (also called detrital or clastic)

1. Conglomerate
2. Breccia
3. Quartz sandstone
4. Arkose
5. Lithic sandstone (also called litharenite or graywacke)
6. Siltstone
7. Shale
8. Claystone
9. Mudstone
10. Mudshale

Chemical and Biochemical

1. Evaporites
   a. Rock salt
   b. Rock gypsum
   c. Travertine
2. Carbonates
   a. Limestone
   b. Dolostone
3. Siliceous Rocks
   a. Diatomite
   b. Chert
4. Sedimentary Ironstones
   a. Oolitic hematite or oolitic ironstone
   b. Iron oxide concretions
   c. Banded iron formations

Organic (Coals)

1. Peat
2. Lignite
3. Bituminous coal
4. Anthracite coal (metamorphosed coal)

METAMORPHIC ROCKS

Foliated

1. Slate
2. Phyllite
3. Schist
4. Gneiss

Nonfoliated or Weakly Foliated

1. Marble
2. Quartzite
3. Greenstone
4. Metabasalt
5. Amphibolite
6. Anthracite coal

Rocks and Minerals Exercises

PRE-LAB EXERCISES

Do these exercises before you come into lab, and have them ready to turn in.

1. List three minerals in this lab that can scratch glass.

   _____

   _____

   _____

2. List two minerals in this lab that fizz in hydrochloric acid. Tell which one fizzes only if it is scratched.

   _____

   _____

   The one that fizzes only when scratched is _____

3. List three minerals in this lab that are softer than your fingernail.

   _____

   _____

   _____

4. Which mineral in this lab has a metallic luster?

   _____

5. Explain how to tell the difference between quartz and feldspar.

   _____

6. For each of these igneous rocks, indicate whether it cooled quickly or slowly.

Igneous Rock	Quick Cooling	Slow Cooling
Granite		
Diorite		
Obsidian		
Rhyolite		
Gabbro		
Andesite		
Basalt		
Pumice		
Peridotite		

7. List which igneous rock(s) contain the following minerals.

Minerals	Igneous Rock(s)
Quartz	
Potassium feldspar	
Plagioclase feldspar	
Biotite	
Olivine	
Minor amounts of pyroxene	
Plagioclase feldspar	
Amphibole	
Pyroxene	

8. Explain the difference between shale and claystone.

9. List the three types of sandstone and name the dominant mineral or type of grain in each.

Type of Sandstone	Dominant Mineral or Type of Grain

10. Some rocks can be identified using hydrochloric acid. Which sedimentary and metamorphic rock(s) fizz in acid?

Sedimentary Rocks that Fizz in Acid	Metamorphic Rock that Fizzes in Acid

11. For each of these metamorphic rocks, check the box to indicate whether it is foliated or nonfoliated.

Metamorphic Rock	Foliated	Nonfoliated
Marble		
Schist		
Slate		

Metamorphic Rock	Foliated	Nonfoliated
Quartzite		
Gneiss		
Metabasalt		
Phyllite		

12. Tell whether the following rocks are igneous, sedimentary, or metamorphic.

Rock	Igneous	Sedimentary	Metamorphic
Granite			
Basalt			
Shale			
Limestone			
Marble			
Quartz sandstone			
Quartzite			
Breccia			
Slate			
Andesite			
Arkose			
Greenstone			
Amphibolite			
Conglomerate			
Graywacke			
Pumice			
Chert			
Gabbro			
Rock gypsum			

13. Examine the rock in the figure below (scale in centimeters) and answer the following questions.

   a. Is this rock igneous, sedimentary, or metamorphic? _____

   b. Describe the texture of this rock. Use three texture terms from this lab.

   _____

   c. Identify the green mineral with the glassy luster. _____

d. Is this rock sialic, intermediate, mafic, or ultramafic? _____

e. Where on Earth might you expect this type of rock to form? _____

f. If this rock were to be metamorphosed, what type of metamorphic rock would it form? _____

Pamela Gore

LAB EXERCISES

Instructions for Lab 2—Rock and Mineral Laboratory

Examine the mineral and rock specimens provided by your instructor. You will need to have the following materials available to assist you with examining and testing these samples: your fingernail, piece of copper, steel nail or dissecting needle, piece of glass, magnet, dilute hydrochloric acid, 10X hand lens or dissecting microscope, ruler with millimeter scale or grain size comparator chart.

Instructions for Filling in the Mineral Identification Table
Column 1: Color

Write down the color or colors of the mineral specimens.

Column 2: Hardness

Use your fingernail, copper, steel nail or dissecting needle, and glass to test the hardness of the mineral by scratching the mineral across the item, or using the item to scratch the mineral. You should be able to bracket the hardness of the mineral with respect to the various items.

Examples:

If the mineral scratches glass, write **H > glass**. The hardness is 6 or greater on Mohs scale.

If the mineral will scratch a nail but is softer than glass, write **nail < H < glass**. This is hardness 5 on Mohs scale.

If the mineral will scratch copper but is softer than a nail, write **copper < H < nail**. This is hardness 4 on Mohs scale.

If the mineral is harder than your fingernail but softer than copper, write **fingernail < H < copper**. This is hardness 3 on Mohs scale.

If you can scratch the mineral with your fingernail, write **H < fingernail**. This is hardness 1 or 2 on Mohs scale.

Column 3: Luster

Determine whether the mineral is **metallic** (is it a metal?) or **non-metallic**. If it is non-metallic, you may further categorize its luster as **glassy** (shiny) or **dull**, or use some of the additional luster terms in Box 2.2.

Column 4: Cleavage

Does the mineral break to produce parallel sets of flat sides? If so, it has cleavage.

> If the mineral breaks into flat sheets, write **1 direction**.
> If the mineral breaks into cubes, write **cubic** or **3 directions at 90°**.
> If the mineral breaks into rhombohedrons (like a deformed cube with angles other than 90 degrees), write **rhombohedral** or **3 directions ≠ 90°**.
> If the mineral has 2 sets of parallel sides, but irregular breakage in other places, write **2 directions**. If you can determine the angles between these two sets of cleavage planes, **write 2 directions at 90°** or **2 directions ≠ 90°**.

Column 5: Reaction to Hydrochloric Acid

Use hydrochloric acid (HCl) to test the sample to determine whether calcite or dolomite is present. Calcite fizzes readily in acid. Dolomite fizzes only if scratched or powdered.

> Test the mineral with 1 drop of dilute HCl and examine it to see whether it reacts and effervesces (or fizzes), producing bubbles of carbon dioxide gas.
> If it fizzes, write **yes**.
> If it does not fizz, write **no**.
> If it fizzes only when scratched and powdered, using a nail or dissecting needle, write **only if scratched**.
> Blot up all acid drops immediately with a paper towel.
> It is not necessary to put HCl on all specimens. You only need to do this if you suspect that the mineral may be calcite or dolomite, both of which have rhombohedral cleavage.

Column 6: Other comments

List any other distinctive features about the sample, such as whether it is magnetic, has conchoidal fracture, or crystals.

Column 7: Mineral Name

Use Tables 2.3 and 2.4 to help you identify the mineral samples. Write the mineral name in this column.

Instructions for Filling in the Igneous Rock Identification Table
Column 1: Texture

Describe the texture of the igneous rock. Use one or more of the following terms: glassy, vesicular, aphanitic, phaneritic, pegmatitic, porphyritic.

Column 2: Minerals Present

Using your 10× hand lens or a dissecting microscope, carefully examine and identify the minerals in the rock. The most common minerals in igneous rocks include quartz, feldspar, biotite, muscovite, amphibole, pyroxene, and olivine. If the rock is glassy, vesicular or aphanitic, you may need to indicate "none visible".

Column 3: Composition Group

Describe the composition of the igneous rock. Use one of the following terms: sialic, intermediate, mafic or ultramafic.

Column 4: Igneous Rock Name

Use Table 2.5 to help you identify the igneous rock using the texture and composition group. Write the igneous rock name in this column.

Instructions for Filling in the Sedimentary Rock Identification Table

Column 1: Texture

If grains are visible in the sample, tell whether they are gravel, sand, silt, or clay sized. You can use a ruler or grain size comparator to determine the grain size.

If grains are not visible (for example in chemical, biochemical, or organic sedimentary rocks), write "**grains not visible**", or put a dash (–) in the texture column.

Column 2: Minerals Present

Use your 10× hand lens or a dissecting microscope to carefully examine and identify the minerals in the rock. The most common minerals in sedimentary rocks include calcite, dolomite, gypsum, halite, quartz, feldspar, clay minerals and iron-bearing minerals. If the sample is coal, it will not have any minerals, so put a dash (--) in the mineral column.

Column 3: Reaction to Hydrochloric Acid

Use hydrochloric acid (HCl) to test the sample to determine whether calcite or dolomite is present. Calcite fizzes readily in acid. Dolomite fizzes only if scratched or powdered.

> Test the rock with 1 drop of dilute HCl and examine it to see whether it reacts and effervesces (or fizzes), producing bubbles of carbon dioxide gas.
> If it fizzes, write **yes**.
> If it does not fizz, write **no**.
> If it fizzes only when scratched and powdered, using a nail or dissecting needle, write **only if scratched**.
> Blot up all acid drops immediately with a paper towel.

Column 4: Sedimentary Rock Name

Use Chart 2.5, along with the photos and descriptions of the other sedimentary rocks, to identify the specimens provided.

Instructions for Filling in the Metamorphic Rock Identification Table

Column 1: Texture

Look for layering or alignment of minerals to determine whether the rock is **foliated** or **non-foliated**.

Column 2: Minerals present

Use your 10× hand lens or a dissecting microscope to carefully examine and identify the minerals in the rock. The most common minerals in metamorphic rocks include muscovite, biotite, chlorite, amphibole, quartz, feldspar, calcite, garnet, staurolite, and kyanite.

Column 3: Reaction to Hydrochloric Acid

Use hydrochloric acid (HCl) to test the sample to determine whether calcite or dolomite is present. If it fizzes, write **yes**. If it does not fizz, write **no**.

> If it fizzes only when scratched and powdered, using a nail or dissecting needle, write **only if scratched**.
> Blot up all acid drops immediately with a paper towel.

Column 4: Metamorphic Rock Name

Use the photos and descriptions of the metamorphic rocks to identify the specimens provided.

1. Identify the minerals provided.

Number	Color	Hardness	Luster	Cleavage	Reaction to Acid	Other Comments	Mineral Name
1							
2							
3							
4							
5							
6							
7							
8							
9							
10							
11							
12							

2. Identify the igneous rocks provided.

Number	Texture	Minerals Present	Composition Group	Igneous Rock Name
13				
14				
15				
16				
17				

3. Identify the sedimentary rocks provided.

Number	Texture	Minerals Present	Reaction to Hydrochloric Acid	Sedimentary Rock Name
18				
19				
20				
21				
22				

4. Identify the metamorphic rocks provided.

Number	Texture	Minerals Present	Reaction to Hydrochloric Acid	Metamorphic Rock Name
23				
24				
25				
26				
27				

OPTIONAL ACTIVITY

Take a field trip around the campus (or your local area) to examine the types of rock used as building and paving stone, to see local rock outcrops, or to see a stream or beach to look at sediment or go to a cemetery to examine rock types used in monuments. Rates of rock weathering can be estimated by examining engravings on various types of monuments, particularly those that have dates.

Rock Weathering and Interpretation of Sediments

This lab introduces the products of rock weathering. Weathering is important because it is the process through which rocks are broken down and sediment is formed. Sediment is loose particulate material that becomes cemented and compacted to form sedimentary rocks.

TYPES OF WEATHERING

There are three major types of weathering:

- Physical
- Chemical
- Biological

Physical weathering breaks rocks down into smaller pieces. Types of physical weathering include frost wedging, exfoliation, and thermal expansion.

Chemical weathering breaks rocks down chemically by adding or removing chemical elements, thereby altering the minerals and making new materials. Chemical weathering proceeds via chemical reactions, most of which involve water. Types of chemical weathering include dissolution, hydrolysis, and oxidation.

Biological weathering is the breakdown of rock caused by the action of living organisms, including plants, burrowing animals, and lichens (crusty, rubbery, light green organic material that grows in patches on rocks). **Lichens** are fungi and algae, living together in a symbiotic relationship (Figure 3.1). Lichens can live on bare rock (in addition to trees, human-made objects, and even certain animals), and they break down rocks by secreting acids and other chemicals. The fungal part of the association secretes the acids, which react to dissolve the minerals, which are then used by the algae. Later, water seeps into crevices etched by the acid and assists in the breakdown through freezing (frost wedging) and chemical weathering.

Types of Physical Weathering

Frost Wedging
Water expands in volume about 9% when it freezes. When water fills a crack in rock, it enlarges the crack by freezing and expanding, and this helps wedge rocks apart into angular fragments.

Figure 3.1 Several varieties of green lichen on a rock. For scale, the stick is about 1 cm in diameter.

Exfoliation

The bedrock breaks into flat sheets along joints roughly parallel to the ground surface. This phenomenon is caused by the expansion of rock when the pressure of overlying rock is removed by erosion. It is also sometimes called *unloading*. The reason that it happens is a physical adjustment for rocks that form at great depths in the crust but then find themselves at the surface of the crust, where the pressures are much, much lower. Given this release of pressure, they expand outward and fail or break in a brittle fashion.

Exfoliation of granite at Stone Mountain, Georgia, has produced a rounded mountain (Figure 3.2). It is calculated that at the time the granite body cooled, the land in this area stood about 10,000 feet higher than at present. Over the past 325 million years, 10,000 feet of formerly overlying rock has been eroded away.

Thermal Expansion

Heat causes expansion of rock; cooling causes contraction. Different minerals expand and contract at different rates, causing stresses along mineral boundaries. Other stresses are induced between the warm outer surface of a rock and its cool interior. Repeated daily heating and cooling of rock causes the rock to break down, very much the same way that repeated bending of a paper clip eventually causes it to break.

A. The rounded shape of Stone Mountain, Georgia, is result of exfoliation.

B. Active exfoliation at Stone Mountain, Georgia.

Figure 3.2

Types of Chemical Weathering

Dissolution

Dissolution alters rocks by removing soluble minerals (Figure 3.3). Minerals such as halite, gypsum, and calcite are soluble in water. They can dissolve (go into solution), especially in water that is slightly acidic. When these minerals react with water, ions (such as Ca^{2+} and Na^+) are released. The ions are carried as dissolved load by rivers flowing to lakes or to the ocean. As lake water or seawater evaporates, the dissolved minerals precipitate out. They crystallize as solid minerals. (Examples are halite, gypsum, and calcite. Minerals that form from the evaporation of water are called **evaporites**.) Minerals can also precipitate or crystallize from ground water in and around springs (particularly hot springs), and in caves, forming the rock called **travertine**.

Figure 3.3   Dissolution of a limestone carving of a shield above a doorway, probably caused by acid rain, in Castillo de San Marcos, St. Augustine, Florida, built between 1672 and 1695. The building is constructed of coquina, which has experienced dissolution over about 325 years of weathering.

Hydrolysis

Hydrolysis is the process by which feldspar and some other aluminum-bearing silicate minerals are weathered to form clay. Literally, the word *hydrolysis* means "to break with water." For example, potassium feldspar weathers to form the clay mineral kaolinite (Table 3.1).

In humid climates, most of the feldspar in rocks such as granite will weather through hydrolysis to form clay. Nearly all of the minerals in the common rocks of Earth's crust weather to form clay (with the exception of quartz). Because of this, clays are the minerals that make up nearly half of the sedimentary rocks on Earth.

TABLE 3.1   CHEMICAL REACTION OF FELDSPAR WEATHERING BY HYDROLYSIS TO FORM KAOLINITE

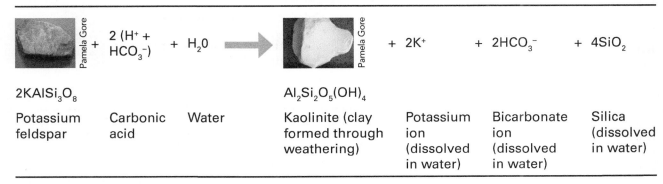

$$+ \ 2(H^+ + HCO_3^-) \ + \ H_2O \ \longrightarrow \ \ + \ 2K^+ \ \ + \ 2HCO_3^- \ \ + \ 4SiO_2$$

$2KAlSi_3O_8$			$Al_2Si_2O_5(OH)_4$			
Potassium feldspar	Carbonic acid	Water	Kaolinite (clay formed through weathering)	Potassium ion (dissolved in water)	Bicarbonate ion (dissolved in water)	Silica (dissolved in water)

Oxidation

Oxidation is the process by which iron-bearing minerals weather to produce iron oxides (or rust) (Figure 3.4). Iron-bearing silicate minerals that also contain aluminum (such as pyroxene, amphibole, and biotite) undergo both oxidation and hydrolysis, forming both iron oxides and clays. Iron-bearing aluminosilicate minerals weather to form red clay-rich soils, as well as lateritic soils (laterite) formed in more tropical areas.

Figure 3.4 Mafic igneous rock, diabase, with weathering rind showing oxidation (orange coloration), from a Jurassic diabase dike at Stone Mountain, Georgia.

MINERAL STABILITY IN THE WEATHERING ENVIRONMENT

Some minerals weather more quickly than others. A few minerals are readily soluble in slightly acidic water, whereas others weather to produce clay. Still others are very resistant to weathering, and they persist for a long time without alteration. One of the controls on the weathering of minerals is the temperature at which the minerals originally formed when they crystallized from magma or lava.

Minerals that formed at high temperatures and pressures are least stable in the weathering environment, and weather most quickly. This is because they are farther from their zone of stability, or the conditions under which they formed. On the other hand, *minerals that formed at lower temperatures and pressures are most stable under weathering conditions.* They are closer to equilibrium with the conditions that they encounter at Earth's surface.

Goldich Stability Series

The order in which minerals tend to weather is related to the temperature at which they crystallized. **Bowen's reaction series** describes the order in which minerals crystallize from magma. A similar ordering of minerals, related to their weathering rates, is called the **Goldich stability series** (Table 3.2). The order of mineral stability in the weathering environment is *the same order* as Bowen's reaction series, but with weathering it is called the Goldich stability series.

TABLE 3.2   GOLDICH STABILITY SERIES

	Mafic Minerals	Feldspars
Least stable (high-temperature minerals)	Olivine	
	Pyroxene	Calcium (Ca) plagioclase feldspar
Increasing mineral stability	Amphibole	Sodium (Na) plagioclase feldspar Potassium (K) feldspar
	Biotite	
	Other Common Rock-Forming Minerals	
Most stable (low-temperature minerals)	Muscovite	
	Quartz	

What Happens When Granite Is Weathered

Unweathered granite contains the following minerals (Figure 3.5):

- Potassium (K) feldspar (pink or white)
- Sodium (Na) plagioclase feldspar (white)
- Quartz (gray)
- Small amounts of biotite and/or amphibole (black)
- Small amounts of muscovite (not shown in the figure)

Figure 3.5   Pikes Peak Granite, Colorado. (Scale in centimeters.)

Here is what happens to each of the mineral constituents in granite under warm, humid weathering conditions (Figure 3.6):

The **feldspars** undergo **hydrolysis** to form kaolinite (**clay**) and sodium and potassium ions, which go into solution.

The **sodium and potassium ions** in the feldspar are removed through leaching and are carried **in solution** in running water.

The **biotite and/or amphibole** undergo **hydrolysis** to form **clay**, *and* they undergo **oxidation** to form **iron oxides** (red to brown).

The **quartz** (and **muscovite**, if present) remains as **residual mineral**, because quartz and muscovite are very **resistant to weathering** and are stable at Earth surface conditions.

Figure 3.6   Granite *(left)* weathers to granite saprolite *(right)*. Mineral grains are about 1 to 2 mm in diameter. (Stone Mountain Granite, Georgia.) (Scale in centimeters.)

Under warm, humid conditions, the granite bedrock weathers in place until the feldspars alter to soft clay. The weathered rock is called **saprolite**, a term meaning "rotten rock" (Figure 3.7). In areas of the temperate southeastern United States that are underlain by granite (and other igneous and metamorphic rocks), a thick soil zone of weathered rock or saprolite has developed. Where the bedrock contained iron-bearing minerals (such as biotite, amphibole, or pyroxene) that weathered to iron oxides, the saprolite has been stained a deep red. (Even a little bit of iron oxide makes the rock a deep red, in much the same way that one red sock in a load of laundry can stain all of the clothes red.)

A. Excavation into gneiss that has been weathered to saprolite. Note that the foliation in the saprolite dips or slopes down to the right. (Decatur, Georgia.)

B. Close-up of **saprolite**. Feldspar-rich layers in the gneiss are weathered by hydrolysis to white kaolinite clay, and layers that contained iron-bearing minerals like biotite and amphibole are weathered by oxidation and hydrolysis to reddish-brown clay. This specimen is about 15 cm wide.

Figure 3.7   Saprolite.

What Happens after the Rock Has Been Weathered to Saprolite

After the rock has been weathered to saprolite, the follow events occur:

The **clays** are eroded and transported by running water to the sea. Clay is fine-grained and remains suspended in the water column. The clay might ultimately be deposited in deep **quiet water** far from shore.

As the soft clay is removed, the unweathered, residual quartz grains are released from the saprolite by erosion. If the quartz in the granite is sand-sized, it becomes **quartz sand**. Larger grains of quartz may be transported as **gravel** or broken into sand-sized pieces. The quartz sand is ultimately transported to the sea, where it accumulates to form **beaches**.

The **dissolved ions** (sodium and potassium) are transported by rivers to the sea and become part of the **salts** in the sea.

In addition to granite, many other igneous rocks and metamorphic rocks also weather to form saprolite under warm, humid conditions.

CHARACTERISTICS OF SEDIMENT

Terrigenous sediment is derived from the weathering of preexisting rocks. (It is also called **clastic** or **siliciclastic** or **detrital** sediment.) The textures and mineralogy of the rocks in the source area control the grain size and composition of the resulting sediment. Texture refers to the size and shape of the grains in sediment.

Describing the Texture of Sediment by Grain Size and Sorting

Texture refers to the size and shape of the grains in a sedimentary deposit. Sediment can be separated into four main groups based on grain size. These four size groups are **gravel, sand, silt,** and **clay.** Some of these groups (gravel and sand) can be further subdivided. The sediment grain size scale is known as the **Wentworth scale** (sometimes called the Udden–Wentworth scale) (Table 3.3).

TABLE 3.3   WENTWORTH SCALE OF SEDIMENT GRAIN SIZE

	Particle Name	Particle Diameter
Gravel	Boulders	>256 mm
	Cobbles	64–256 mm
	Pebbles	4–64 mm
	Granules	2–4 mm
Sand	Very coarse sand	1–2 mm
	Coarse sand	0.5–1 mm
	Medium sand	0.25–0.5 mm
	Fine sand	0.125–0.25 mm
	Very fine sand	0.0625–0.125 mm
Silt		1/256–1/16 mm (or 0.004–0.0625 mm)
Clay		<1/256 mm (or <0.004 mm)

Gravel forms through physical weathering of rock. A piece of gravel is usually a rock fragment composed of more than one mineral. Sometimes a piece of gravel is a single mineral, most commonly quartz. This is because quartz is sometimes present as veins, which may be several inches wide (or more), thus producing gravel-sized clasts.

Sand forms through the breakdown and disintegration of rocks that have sand-sized (1/16–2 mm) grains, such as granite and gneiss. Sand-sized grains can also be produced when larger grains collide and break off smaller pieces. In humid climates, quartz sand grains are released from granite after the feldspar grains alter to clay by chemical weathering (hydrolysis). In more-arid areas, granite breaks down by physical weathering (such as frost wedging), releasing feldspar and quartz grains.

Silt originates from the chipping of coarser grains during sediment transport or from the disintegration of fine-grained crystalline rocks (such as slates and phyllites).

Clay originates primarily through chemical weathering of feldspars and other aluminosilicate minerals (those that contain aluminum and silica). The term *clay* refers to a particular size of sediment particle, which could be a quartz grain, a clay mineral flake, or some other very small mineral fragment. The term *clay* is also used to refer to a group of minerals. There are a number of clay minerals, including **kaolinite** (the white clay used for coatings on paper and for additives to rubber), **illite** (which contains potassium), and **montmorillonite** or **smectite**. Montmorillonite is a group of clays that can absorb large amounts of water, and as a result these clays are commonly referred to as "swelling clays"; kitty litter is one example.

Sorting

Sorting refers to the range in grain sizes in a sediment or sedimentary rock. In sediment (or rock) that is **well-sorted**, most of the grains are roughly the same size; the biggest grain in the sample is about ten times the size of the smallest grain in the sample. A **poorly sorted** sediment or rock has a wide range of grain sizes; the biggest grain in the sample is thousands or tens of thousands of times larger than the smallest grain in the sample). Sorting can be estimated using a visual comparison chart like the one in Figure 3.8.

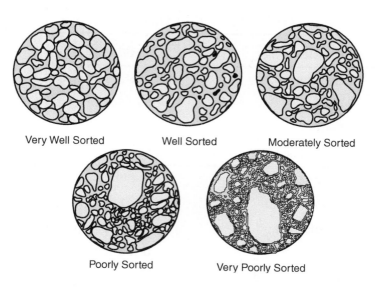

Very Well Sorted   Well Sorted   Moderately Sorted

Poorly Sorted   Very Poorly Sorted

Figure 3.8   Visual comparison chart illustrating sorting in sediment or sedimentary rock. Used with permission of Department of Earth, Ocean and Atmospheric Sciences, The University of British Columbia.

Describing the Texture of Sediment by Grain Shape

The shape of grains of sediment is described in terms of its roundness and sphericity. Although these words seem vaguely similar, they actually refer to quite different properties.

   Roundness is a measure of the sharpness or roundness of the corners of a sedimentary particle. Roundness is determined by comparing the sand grains with a visual comparison chart (Figure 3.9).

Angular   Subangular Subrounded   Rounded   Well rounded

Figure 3.9   Visual comparison chart illustrating rounding in sediment or sedimentary rock. Used with permission of Department of Earth, Ocean and Atmospheric Sciences, The University of British Columbia.

As sediment is transported, it undergoes **abrasion** by coming into contact with the stream bottom, the sea floor, or other grains of sediment. The abrasion tends to round off the sharp edges or corners. Rounding is also related to the size of the grains. Boulders tend to round much more quickly than sand grains because they strike each other with much greater force. Chemical weathering also helps enhance the rounding of sediment through the greater surface area available for chemical attack on the corners and edges (Figure 3.10).

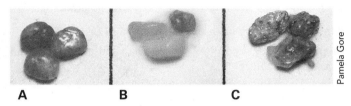

A **B** **C**

Figure 3.10 Roundness of sand grains.
A. Well-rounded. **B.** Subrounded. **C.** Subangular.

Grains of sediment are three dimensional. **Sphericity** refers to equal dimensions. Is the sediment particle elongate (one dimension longer than the other two), flat or sheetlike (one dimension much smaller than the other two dimensions), or spherical (its three dimensions roughly the same length)? Sphericity can be described as high or low. According to this definition, a ball has high sphericity, but so does a cube (high sphericity, but low roundness). In contrast, a submarine sandwich or a hot dog has low sphericity but high roundness. A shoebox has both low sphericity and low roundness. Sand grains can have high or low sphericity. Some minerals produce elongated or flattened grains, depending primarily on the original crystal shape and cleavage.

Be careful not to confuse rounding with sphericity. A well-rounded grain might or might not resemble a sphere, and a spherical grain might or might not be well-rounded (Table 3.4).

TABLE 3.4 ROUNDNESS VERSUS SPHERICITY

Roundness	Sphericity	
	Low Sphericity	High Sphericity
High roundness	⬭	⬤
Low roundness	▯	◻

Textural Maturity

Textural maturity is a concept that proposes that as sediments experience the input of mechanical energy (the abrasive and sorting action of waves and currents), they pass through a series of four stages.

Stage 1. Immature: Sediment containing mud (clay and/or silt)

Stage 2. Submature: Poorly sorted sediment with no mud

Stage 3. Mature: Well-sorted sediment with no mud

Stage 4. Supermature: Well-sorted and rounded sediment with no mud

Three steps are involved in moving sediments through these stages (Table 3.5):

Winnowing or washing out of clay and other fine-grained sediment makes an immature sediment become submature.

Sorting of grain sizes makes a submature sediment become mature.

Rounding makes a mature sediment become supermature.

TABLE 3.5   STAGES OF TEXTURAL MATURITY

	Winnowing		*Sorting*		*Rounding*	
1 **Immature**	→	**2** **Submature**	→	**3** **Mature**	→	**4** **Supermature**
Muddy sediment		Poorly sorted sediment		Well-sorted sediment		Well-sorted and well-rounded sediment

---

INTERPRETING THE TEXTURE OF SANDS

Texture is an indicator of **energy levels** in the environment of deposition (the place where sediment accumulates, perhaps a beach, a riverbed, a lake, or a delta). Moving water (such as waves or currents) is considered a **high-energy** environment. Quiet water or still water (water without waves or currents) is considered a **low-energy** environment. Deep-water environments commonly have quiet water, because wave motion is restricted to the upper part of the water column.

How do you determine energy levels in the depositional environment from looking at sediment?

Interpreting Grain Size

Coarse-grained sediments (sand, gravel) indicate high-energy environments. A large amount of energy is required to transport gravel-sized clasts, and moving water is required to transport sand.

Fine-grained sediments (clay or silt) indicate low-energy environments. There is insufficient energy to bring larger clasts into the environment. Also, if the water were moving, the clay would not be able to settle out and be deposited on the bottom.

Interpreting Sorting

Well-sorted grains indicate that the sediment was probably transported for a long time in a fairly high energy environment (waves or currents). The finer grains were probably washed or **winnowed** away.

Poorly sorted grains indicate that the sediment has not been transported very far from the source area. It also suggests fluctuating energy levels and a fairly short time in the depositional environment.

- Good sorting implies consistent energy (washing).
- Poor sorting implies inconsistent energy (dumping).

Interpreting Grain Shape

A well-rounded sand grain indicates that the sediment has been transported far from the original source area and that it has been in the depositional environment for a long time. The environment of deposition is also a factor in sand grain roundness. Sands from desert environments tend to be more rounded than sands from beaches. Angular sand grains have probably only been transported for a short distance from the source area, or they have been in the depositional environment for a short time.

DESCRIBING THE MINERALOGIC COMPOSITION OF SANDS

It is important to keep in mind that *sand* is a *texture* term, *not a composition* term. Sand can be composed of any types of sand-sized mineral or rock-fragment grains. All that makes them sand is their size.

The minerals in sands (and in sandstones) can be identified using a microscope (or a hand lens if a microscope is not available). Identifying the minerals present is important because sandstones are classified based on the composition of their constituent grains.

Terrigenous Sands

Three components are considered when naming terrigenous sandstones (Figure 3.11):

1. **quartz** grains,
2. **feldspar** grains, and;
3. **fine-grained rock fragment** grains.

Possibilities for fine-grained rock fragment grains include shale, slate, phyllite, basalt, rhyolite, andesite, chert, and possibly schist. Limestones would not usually be included because they dissolve so readily.

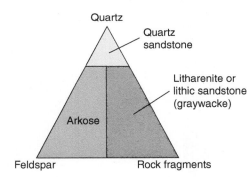

Figure 3.11 Simplified sandstone classification diagram.

The three major types of sandstone are:

1. **Quartz sandstone** (also called **quartz arenite**), which is dominated by **quartz**
2. **Arkose**, which is dominated by **feldspar**
3. **Litharenite** or **lithic sandstone** (commonly but imprecisely called **graywacke**), which is dominated by **rock fragment grains**.

Each type of sand or sandstone implies something about depositional history. **Quartz sandstone** implies a long time in the depositional basin. **Arkose** implies a short time in the depositional basin because feldspar typically weathers quickly to clay. Arkose also implies rapid erosion, arid climate, tectonic activity, and steep slopes. **Litharenite (graywacke)** implies rapid erosion and temperate or arid (not humid) climate.

Other minerals may also be present in sands and sandstones. In fact, in some areas, sands may be composed almost entirely of minerals other than quartz and feldspar. For example, at White Sands National Monument in New Mexico, the sands are composed of gypsum grains. As another example, on the southern tip of the Big Island of Hawaii, there is a beach that has green sand composed mainly of olivine grains.

Heavy Minerals in Terrigenous Sand

In addition to the major constituents in sand, there is commonly a suite of **heavy minerals** (those with high specific gravity, i.e., greater than 2.85), which can consist of less than 1% of the sand grains to perhaps several percent (or more) (Figure 3.12). Examples of heavy minerals include rutile, ilmenite, tourmaline, zircon, garnet, kyanite, staurolite, apatite, olivine, pyroxene, amphibole, magnetite, ilmenite, hematite, and pyrite. The particular types of heavy minerals present depend on the composition of the rocks in the sediment source area. For example, garnet, kyanite, and staurolite come from metamorphic rocks, whereas olivine, pyroxene, and amphibole generally come from mafic igneous rocks (gabbro and basalt). Heavy minerals are important indicators that can tell us the type of rocks that existed in the sediment source area.

Heavy minerals are economically valuable, although they account for only a small portion of sand deposits. The black sands along the coast of the southeastern United States and inland in the Pleistocene barrier island sand ridges of the Coastal Plain are composed of heavy minerals. One notable heavy mineral deposit is in Trail Ridge, the sand barrier holding back the waters of the Okefenokee Swamp. The Trail Ridge sands contain ilmenite, rutile, and zircon heavy mineral grains derived from the weathering of crystalline rocks in the Piedmont. Ilmenite and rutile contain titanium, an element used as a white pigment in paper, plastics, paint, textiles, cosmetics, toothpaste, and foods (such as the m on M&Ms and the creamy filling of Oreo cookies). Zircon sands are used for high-temperature ceramic and metal casting, such as heat-resistant tiles for spacecraft. Staurolite is used as an abrasive for sandblasting. Although the sand is only about 2% to 3% heavy mineral ore, the Trail Ridge deposit is one of the few places in the world where these minerals are found, and they have been mined in Florida since 1949.

Pamela Gore

Figure 3.12 Streaks of black heavy mineral sands on the beach, with driftwood and cord grass (salt marsh grass) at Tybee Island, Georgia. The area shown is about 2 feet across.

Calcium Carbonate Sand

Sands from warm, shallow, tropical seas, far from land-derived sediments, may be composed almost entirely of microscopic shells (the remains of planktonic organisms) and fragments of marine organisms. These sands are made of calcium carbonate ($CaCO_3$) grains such as calcite and aragonite. Other calcium carbonate grains may be spherical or highly rounded. These grains are known as **oolites** (Figure 3.13). (Calcite and aragonite have the same chemical formula but different crystal structures. For this course, you do not need to be able to tell them apart.) Calcium carbonate sand grains are typically white to tan to pink and are opaque rather than transparent and glassy in appearance, like quartz (Figure 3.14). Grains made of calcite round relatively rapidly because calcite is a soft mineral (only 3 on Mohs hardness scale). Recently broken shells, however, are angular. You are likely to see some carbonate sands in your lab.

Figure 3.13 Oolite sand. (Scale in millimeters.)

Figure 3.14 Carbonate sand, Bermuda. (Scale in millimeters.)

IDENTIFYING MINERALS AND ROCK FRAGMENTS IN SANDS

Table 3.6 is a guide to identifying sand grains under the microscope or hand lens. The most common minerals are listed near the top of the table, and the remainder are in alphabetical order.

TABLE 3.6   IDENTIFYING MINERALS IN SAND GRAINS UNDER MAGNIFICATION

Grain Type	Identifying Features
Quartz	Gray, white, or colorless (may be covered by brownish iron oxide stain); glassy; lacks cleavage; displays conchoidal fracture
Feldspar	Usually white or pink, has cleavage (look for flat surfaces or square corners)
Rock fragments	Commonly dark gray, greenish gray, black, or tan; fine-grained; may be coated with a rusty iron oxide stain
Muscovite	Silvery color, shiny, flat sheets; can look submetallic
Amphibole	Greenish to black, elongated to fibrous
Apatite	Colorless, rounded or elongated; may be confused with quartz, which is much more common
Biotite	Brown, shiny, flat sheets
Garnet	Most commonly clear, pale pink or red, no cleavage
Hematite	Red
Magnetite or ilmenite	Black, opaque; magnetite is magnetic
Olivine	Olive green, glassy; may be rounded; no cleavage
Pyrite	Brassy gold, metallic
Pyroxene	Gray or greenish to colorless, stubby, angular cleavage fragments
Rutile	Deep red or yellow, can look opaque, generally elongate and well-rounded
Staurolite	Brown to yellow, elongate, may be filled with tiny inclusions to resemble Swiss cheese
Tourmaline	Dark color, elongated with triangular cross-section
Zircon	Colorless, elongated crystals

READING THE RECORD IN THE ROCKS: A SANDSTONE INTERPRETATION GUIDE

One of the goals of historical geology is to try to interpret the depositional conditions of the sedimentary rocks that make up the geologic record. Sandstone textures and mineral compositions may be used to interpret many things about the history of the sand, including source area lithology (rock types upstream from where the sediment was deposited), paleoclimate, tectonic activity, processes acting in the depositional basin, and time duration in the basin. Remember that the **source area** is the land that is weathering and eroding to supply terrigenous debris to the depositional basin. Also remember the difference between **weathering** (*breakdown* of rock by hydrolysis, dissolution, oxidation, exfoliation, frost wedging, or freeze thaw) and **erosion** (*transportation* of particles).

Source Area Lithology

The composition of a sand or sandstone (the particular types of minerals or rock fragments present) gives the key information on the **lithology** (or rock types) in the sediment source area (Table 3.7). Remember that quartz sandstone or quartz arenite is dominated by **quartz** grains, arkose is dominated by **feldspar** grains (usually potassium feldspar), and litharenite or graywacke is dominated by **rock fragment** grains.

TABLE 3.7   INTERPRETING THE SOURCE AREA LITHOLOGY

Composition of Grains in the Sand or Sandstone	Rock Types in the Sediment Source Area
Quartz sand grains	Granite, gneiss, or older sandstones that contain quartz (recycled sandstones)
Feldspar sand grains	Granite or gneiss
Rock fragment sand grains	Fine-grained rocks, such as shale, slate, phyllite, basalt, rhyolite, andesite, chert, and possibly schist

As noted earlier, the particular suite of heavy minerals present in sand also can indicate a lot about the source area lithology.

Paleoclimate

Paleoclimate refers to the climate that existed in the source area when the rocks were weathering to produce sediment. We are particularly concerned with weathering rates and processes here. Remember that in *humid* climates, feldspar weathers to **clay** by **hydrolysis**. Other minerals, such as olivine, pyroxene, and amphibole, also weather to clay with associated iron oxides. Table 3.8 shows the likely paleoclimate of an area when certain minerals or types of rock fragments are present in sand or sandstone.

If **rock fragments** are present in your sand, it helps to know what lithology they are. If you cannot identify them, a good compromise answer is *temperate climate*.

Tectonic Activity in the Source Area

We are basically classifying tectonic activity here as **active** or **passive** (Table 3.9). For a good model, consider the west coast of the United States as tectonically active: steep slopes, mountains close to the sea, significant number of earthquakes, tectonic uplift, and occasional volcanic activity. On the other hand, consider the east coast of the United States to be tectonically passive: broad, flat coastal plain; few or no earthquakes; little or no uplift; and no volcanic activity.

TABLE 3.8   INTERPRETING THE PALEOCLIMATE

Composition of Grains in the Sand or Sandstone	Paleoclimate Interpretation	Explanation
Quartz sand grains	Humid	Because the feldspars weathered away to clay
Feldspar sand grains	Arid	The presence of feldspar can also indicate that erosion rates were very rapid and tectonic activity was extremely high, with significant uplift that resulted in steep slopes
Rock fragments of limestone or basalt*	Arid	These rocks dissolve or oxidize rapidly in humid climates
Rock fragments of shale, slate, or chert*	Temperate to humid	These rocks weather slowly in humid climates and are relatively stable

*Remember to use Bowen's reaction series and the Goldich stability series to determine the relative stability of minerals in the weathering environment.

TABLE 3.9   INTERPRETING TECTONIC ACTIVITY IN THE SOURCE AREA

Composition of Grains in the Sand or Sandstone	Tectonic Activity
Quartz sand grains	Tectonically passive, low tectonic activity
Feldspar sand grains	Tectonically active, high tectonic activity

Tectonic activity also influences sorting, time duration in the depositional environment, and, to some extent, **compositional maturity** (percentage of quartz in the sediment). High tectonic activity might produce rapid dumping of sediments into the basin with little or no time for sorting. Low tectonic activity means little uplift, low erosion rates, and therefore little sediment supplied to the basin. What sediment that is there is likely to wash around for a long time and become well-sorted and rounded, and grains other than quartz are likely to be destroyed (by abrasion or chemical weathering).

Processes Acting in the Depositional Basin

"Processes acting in the depositional basin" refers to **energy levels** (high vs. low) and **consistency** of energy or movement of water (or wind) (Table 3.10). The texture of the sand or sandstone gives the key information—in particular, the grain size and sorting.

Time Duration in the Depositional Environment

Both mineralogy and texture can be used to determine the relative amount of time that the sediment spent in the depositional environment before burial and lithification (compaction and cementation, or recrystallization) (Table 3.11).

TABLE 3.10   INTERPRETING ENERGY LEVELS IN THE DEPOSITIONAL ENVIRONMENT

Texture of Sand or Sandstone	Energy Levels
Grain Size	
Coarse grained	High energy
Fine grained	Low energy
Sorting	
Well-sorted	Consistent, fairly high energy levels
Poorly sorted	Inconsistent energy levels: rapid dumping, which might involve short episodes of high energy followed by low-energy conditions

Mineralogical Indicators of Time in the Depositional Environment

Because quartz grains are more resistant to abrasion and chemical breakdown than feldspar or rock fragments are, their presence or absence provides clues to the relative amount of time spent in the depositional environment before burial. Sand with abundant quartz grains suggests a long time in the depositional environment. Sand with abundant feldspar or rock fragment grains suggests a short time in the depositional environment before burial.

Textural Indicators of Time in the Depositional Basin

Textural maturity is also useful in interpreting the relative amount of time that the sediment spent washing and rolling around in the depositional environment. **Immature** (muddy) **or submature** (poorly sorted) sediments probably spent only a short time in the basin before burial and lithification. **Mature** (well-sorted) or **supermature** (well-rounded) sediments were probably rolling around the basin for a long time before they were finally buried. **Roundness** is a good clue to a long time in the depositional environment. Rounding of grains takes a long time, and it is more likely to occur in a tectonically passive situation. Desert sands are commonly well-rounded because of the sandblasting process of wind transport. Hence, in an arid desert, it is possible to get a well-rounded (supermature) arkose. Table 3.11 summarizes mineralogical and textural characteristics for interpreting the relative amount of time that sediment has spent in the depositional environment before burial and lithification.

TABLE 3.11   INTERPRETING TIME IN THE DEPOSITIONAL ENVIRONMENT

Mineralogical Indicators	Textural Indicators	Time in the Depositional Environment
Abundant quartz grains	Mature or supermature sediment (well-sorted or well-rounded sediment)	Long time
Abundant feldspar or rock fragment grains	Immature or submature sediment (muddy or poorly sorted sediment)	Short time

Rock Weathering and Interpretation of Sediments Exercises

PRE-LAB EXERCISES

1. For each of the following sedimentary rock descriptions, interpret these five paleoenvironmental factors:

- **source area lithology** (rock type from which it was derived),
- **paleoclimate** (humid, arid, etc.),
- **tectonic activity** (high or low tectonic activity),
- **energy levels** (high or low, consistent or inconsistent energy levels),
- **time in the depositional environment** (long or short).

Sedimentary Rock Descriptions	Paleoenvironmental Factors				
	Source Area Lithology	Paleoclimate	Tectonic Activity	Energy Levels	Time
Quartz sandstone that is well-sorted and well-rounded					
Arkose that is poorly sorted and poorly rounded					
Quartz sandstone that is angular, muddy, and poorly sorted					
Arkose that is well-sorted and well-rounded					
Litharenite that is poorly sorted, has no mud, and has angular grains					

2. Examine the river sand in the figure and answer the following questions.

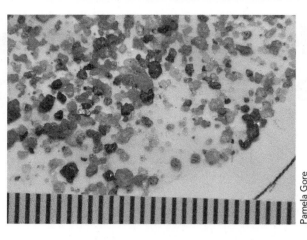

Pamela Gore

River sand, Atlanta, Georgia. (Scale in millimeters.)

a. Using the guide in this lab (Figure 3.8), describe the sorting of this sand. Is this sand very well-sorted, well-sorted, moderately well-sorted, poorly sorted, or very poorly sorted?

_____

b. Using the guide in this lab (Figure 3.9), describe the roundness of this sand. Is this sand well-rounded, rounded, subrounded, subangular, or angular?

_____

c. Has this sand been in the depositional basin (river) for a long time or for a short time?

_____

d. This sand was derived from the weathering of granites and gneisses. What minerals do you expect to be in this sand, and why?

_____

e. Why is this sand brown? What has happened to it?

_____

f. Predict what will happen to this sand as it travels downstream to the Atlantic Ocean.

_____

_____

3. Examine the sand from Colorado in the photo below.

Sand from a dry stream bed at the foot of the Rocky Mountains near Pikes Peak, Colorado. (Scale in millimeters.)

Pamela Gore

   a. The angular pink grains in this sand are what mineral?

   _____

   b. The white or light gray grains in this sand are what mineral?

   _____

   c. There is a dark gray grain on the left side of this image that is not an individual mineral grain. What type of grain is it?

   _____

   d. What type of sandstone would you predict that the sand from Colorado would form?

   _____

   e. What would you interpret about the *source area lithology* from this sand? (Give the *rock type*, not the minerals.)

   _____

   f. What would you interpret about the climate from examining this sand?

   _____

   g. What would you interpret about the time in the depositional basin from this sand?

   _____

   h. Using the guide in this lab, describe the sorting of this sand specimen. Is this sand very well-sorted, well-sorted, moderately well-sorted, poorly sorted, or very poorly sorted?

   _____

   i. Using the guide in this lab, describe the roundness of the grains in this sand specimen. Are they well-rounded, rounded, subrounded, subangular, or angular?

   _____

   j. What would you interpret about the tectonic setting of these sands or their source area?

   _____

   k. What would you interpret about the energy levels of the depositional environment of these sands?

   _____

4. Examine the three beach sands below and answer the following questions. These beach sands come from the southeastern United States. Destin, Florida, is on the Gulf Coast; Kure Beach, North Carolina, and Washington Oaks Gardens State Park, Florida, are on the Atlantic coast. The orange grains are mollusc (clam) shell fragments.

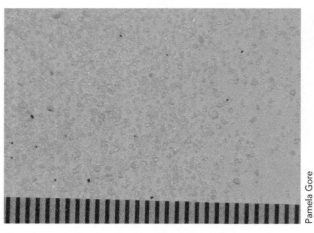

Beach sand, Destin, Florida. (Scale in millimeters.)

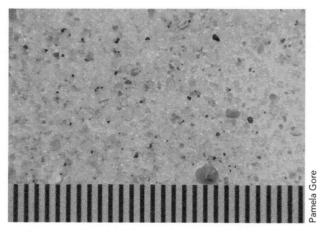

Beach sand, Kure Beach, North Carolina. (Scale in millimeters.)

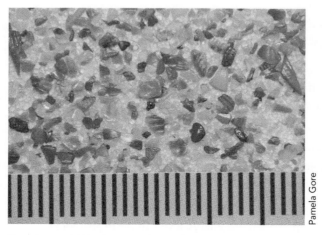

Beach sand, Washington Oaks Gardens State Park, Florida. (Scale in millimeters.)

a. What is the common colorless mineral that all three of these sands have in common?

_____

b. What type of sandstone would you predict that the sands from Destin and Kure Beach would make?

_____

c. The colored grains in the Kure Beach sand include silver or greenish muscovite, black biotite, and other fine-grained black minerals in addition to orange mollusc (clam) shells. What would you interpret about the _source area lithology_ of the Kure Beach sand? (List the _rock type_ or types, not the minerals.)

_____

d. Describe the sorting of each of these sand specimens. Are these sands very well-sorted, well-sorted, moderately well-sorted, poorly sorted, or very poorly sorted?

Sand Specimen	Sorting
Destin, Florida	
Kure Beach, North Carolina	
Washington Oaks Gardens State Park, Florida	

e. How would you interpret the energy levels of the depositional environment of each of these sands?

Sand Specimen	Energy Level
Destin, Florida	
Kure Beach, North Carolina	
Washington Oaks Gardens State Park, Florida	

f. What would you interpret about the climate of the depositional environment from examining these sands?

Sand Specimen	Climate
Destin, Florida	
Kure Beach, North Carolina	
Washington Oaks Gardens, Florida	

g. What would you interpret about the time in the depositional basin from these sands?

_____

Sand Specimen	Time in the Depositional Basin
Destin, Florida	
Kure Beach, North Carolina	
Washington Oaks Gardens, Florida	

h. Which of these three sands has spent the *least time* in the depositional environment?

_____

i. Explain your reasoning for the above answer.

_____

j. What would you interpret about the tectonic setting of these sands or their source area?

_____

k. How do the beach sands differ from the Atlanta river sand?

_____

l. Why do the beach sands differ from the river sand? What processes were involved?

_____

5. Examine the two green sands in the photos below.

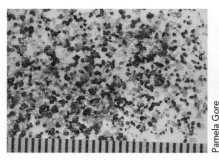
Green sand, Floreana Island,
Galapagos Islands.
(Scale in millimeters.)

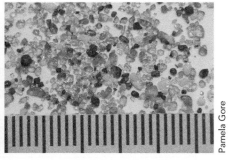
Green sand, Big Island, Hawaii.
(Scale in millimeters.)

a. What is the green mineral in these two sands? _____

b. What do you think the black grains are? _____

c. Interpret the source area lithology. Name *two* types of rock that contain this green mineral. _____

d. Interpret the tectonic activity in the source area. _____

e. What is the stage of textural maturity? _____

f. Interpret the energy levels in the depositional environment.

_____

g. Have these grains been in the depositional environment for a long time or a short time? How did you come to this conclusion?

_____

LAB EXERCISES

The questions in this section require the use of samples of various types of sand. There are also questions dealing with specimens of granite and granite saprolite, if available.

Part 1. Examining and Characterizing Sand Samples

Materials needed:

- Data Table for Sand Samples (below)
- Several different sand samples (may be glued onto cards or loose in petri dishes or vials)
- Microscopes (preferred) or hand lenses
- Sand gauges for determining grain size

Begin by collecting data on each of the sand samples. Examine the sand sample(s) and write your answers to the following questions in the Data Table for Sand Samples, below. The letters of each question correspond to the lettered columns in the Data Table. Space is available in the table for ten sand samples.

A. Where was the sand sample collected? (Read the label.)

B. What color(s) is it? Is it one color or a mixture of many colors?

C. Examine the sand with a microscope (or hand lens, if microscopes are not available).

Is the sand all one mineral, or are there different minerals in it?

What minerals can you identify in the sand?

D. Are any rock fragments present in the sand? If so, can you determine the kinds of rocks?

E. Are there any shells or other remains of organisms in the sand?

F. Look at the size of the sand grains. Compare them with a sand gauge. Write the range of grain sizes in the table. Consider the following questions as you work.

   1. Are all of the grains the same size?

   2. How big is the largest grain of sand? Use the sand gauge to determine grain size.

   3. How big is the smallest grain of sand? Use the sand gauge to determine grain size.

   4. What would you estimate is the average grain size in millimeters of the sand grains?

   5. Think about the sizes of minerals; there is no need to write your answers.

      a. Are some of the minerals typically larger than others?

      b. Which are the larger minerals?

      c. Which are the smaller minerals?

G. Use the guide in this lab (Figure 3.8) to estimate the sorting of the sand sample. Is this sand very well-sorted, well-sorted, moderately well-sorted, poorly sorted, or very poorly sorted?

H. Use the guide in this lab (Figure 3.9) to estimate the roundness of the majority of the grains in this sand sample. Are they well-rounded, rounded, subrounded, subangular, or angular?

DATA TABLE FOR SAND SAMPLES

Sand	A. Location where Collected	B. Color	C. Minerals Present	D. Rock fragments (Yes/No)	E. Shells (Yes/No)	F. Grain Size Range (mm)	G. Sorting	H. Grain Roundness
1								
2								
3								
4								
5								
6								
7								
8								
9								
10								

Part 2. Interpreting Sand Samples

Interpret the following five paleoenvironmental factors for each sand sample that you described (above) by completing the Sand Sample Interpretation table, below. The letters of each question correspond to the lettered columns in the Data Table.

A. Source area lithology (original rock type from which the sand was derived)

B. Paleoclimate (humid, arid, etc.)

C. Tectonic activity (high or low tectonic activity)

D. Energy levels (high or low, consistent or inconsistent energy levels)

E. Time in depositional basin (long or short)

SAND SAMPLE INTERPRETATION TABLE

Sand	Location where Collected*	A. Source Area Lithology	B. Paleoclimate	C. Tectonic Activity	D. Energy Levels	E. Time
1						
2						
3						
4						
5						
6						
7						
8						
9						
10						

*Copy this from the Data for Sand Samples table.

Part 3. Examining Granite and Granite Saprolite

1. Examine a specimen of granite in the lab and describe it.

   a. What color (or colors) is it? _____

   b. Is it hard, or does it crumble easily? _____

2. Describe the texture of the granite. (Use an igneous texture term.)

   _____

3. What is its grain size (in mm)? _____

4. Identify four minerals present in the granite sample.

   _____   _____

   _____   _____

5. Examine the specimen of granite saprolite in the lab and describe it. Do all work over a paper towel or sheet of paper to catch the rock particles

   a. What color is it? _____

   b. Is it hard, or does it crumble easily? _____

6. What is its grain size (in mm)? _____

7. List four minerals you can identify in the saprolite.

   _____   _____

   _____   _____

8. What two minerals are missing from the saprolite compared with the minerals you identified in the granite?

   _____   _____

9. Why are they missing? What happened to them? What weathering processes were involved?

Missing Mineral	Weathering Process(es) Involved

10. Estimate the percentages of each of the following in your saprolite sample. Your percentages must total to 100%.

Mineral	Estimated Percentage
Kaolinite (white clay)	
Quartz	
Muscovite	
Iron oxides (reddish brown)	
Other minerals (if any)	
TOTAL	100%

11. What mineral weathered to form the kaolinite? _____

12. What mineral (or minerals) weathered to form the iron oxides?

   _____

13. What kind of igneous rock contains these minerals? _____

14. Compare and contrast the texture of the saprolite and the texture of the granite.

_____

Part 4. From Saprolite to Sediment

You might do this experiment, or your instructor might demonstrate it for you.

- Crumble some of the saprolite into a jar or empty plastic water bottle.
- Fill the jar or bottle with more than enough water to cover the saprolite.
- Put the lid on and shake the jar vigorously until the saprolite is totally separated into individual grains and no lumps remain.
- Set the jar or bottle down and observe the settling of the sediment, then answer the following questions.

1. How long does it take for the larger grains to settle? _____

2. How long does it take for the finest grains to settle? _____

3. After all of the particles have settled and the water is clear, examine the layering you see. (Or examine the photo below.) Describe the characteristics of each layer you can observe. Use Wentworth Scale terms to describe the grain sizes.

_____

_____

_____

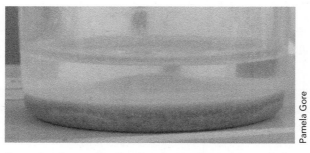

Example of saprolite that has been allowed to settle in a water bottle about 3 inches in diameter.

4. Write a hypothesis about how the various grain sizes settle, to explain your observations.

_____

_____

_____

WEB EXERCISE: THE VIRTUAL SAND COLLECTION

The questions below are adapted with permission of Dr. Dave Douglass, from the Virtual Sand Collection website at Pasadena City College in California: http://faculty.pasadena.edu/dndouglass/sand/SANDHP.htm.

For each question, go to the web links given to see the sand sample as viewed through the microscope. Click on ZOOM IN several times to increase the magnification so that you can get a closer look at individual sand grains under higher magnification with the microscope.

Sand #1: California Beach Sand, San Clemente, California

http://faculty.pasadena.edu/dndouglass/sand/SandPile/SanClem.htm

This light-colored sand appears tan to the naked eye, but most of the grains are color-less under the microscope. The composition of this sand is similar to the composition of granitic rocks found in nearby mountains, namely minerals such as feldspar, quartz, and biotite.

1. Zoom in several times to find and enlarge a grain of the colorless mineral. Describe how the colorless mineral breaks.(Describe the cleavage or fracture.)

   _____

2. Identify the colorless mineral. It is one of the three minerals listed above.

   _____

3. Look for a grain of biotite in the sand. Based on what you learned about mineral weathering in this lab, what two weathering processes affect biotite, and what are the weathering products of each (i.e., what new minerals are formed as biotite is weathered)?

Weathering Process Affecting Biotite	Weathering Product (New Mineral)

4. Because biotite weathers easily, what does the presence of biotite in this sand suggest about how much these grains have been weathered and how far they have been transported? Circle the correct answer below (a or b).

   a. Heavily weathered and/or transported far from the source area.

   b. Not weathered much and not transported far from the source area.

Sand #2: Carbonate Sand from Gun Beach, Guam

http://faculty.pasadena.edu/dndouglass/sand/SandPile/GunBech.htm

Sand #3: Carbonate Sand from O'Carrol's Cove, Ireland

http://faculty.pasadena.edu/dndouglass/sand/SandPile/Ocarrol.htm

Sands #2 and #3 are very different from those on California beaches. The sands are composed mainly of calcium carbonate produced by the activities of living organisms. The sands consist of bits of shells, foraminifera tests, and pieces of other organisms, such as algae or coral. Zoom in several times and examine the microscopic fossils in these two sands. These microfossils are called foraminifera (or forams for short), and their shells are called tests.

5. Sketch a foram test from each sand as it appears under highest magnification.

Gun Beach, Guam	O'Carrol's Cove, Ireland

6. The foraminifera are quite interesting, but they are very different in each sand. Why do you think the forams are so different? *Hint*: Locate Guam and Ireland on a world map.

_____

7. Why do you suppose we do not find carbonate sands dominated by foraminifera on beaches in southern California? *Hint*: Think about the source area lithology, climate, and tectonic setting. How might one or all of these affect California sand to make it different from Sands #2 and #3?

_____

Sand #4: Heavy Mineral Sand from Dockweiler Beach, California

http://faculty.pasadena.edu/dndouglass/sand/SandPile/DocWilr.htm

This is a different kind of sand from southern California called a "heavy mineral sand" because it is largely made up of minerals that have a higher than average density.

8. What colors are the grains in this heavy mineral sand?

_____

9. Zoom in to look at the clear, red sand grains. Using Table 3.6, what is this mineral?

_____

10. Look at the black grains. Click on MAGNET to see what happens when they are placed next to a magnet. Using Table 3.6, what is this mineral?

_____

Sand #5: Volcanic Black Sand from Jokulsa River, Iceland

http://faculty.pasadena.edu/dndouglass/sand/SandPile/Iceland.htm

11. This black sand is very different from the one from Dockweiler Beach. What are three ways these sand grains differ from those at Dockweiler Beach?

_____

_____

_____

12. What additional tests might you want to do on these sand grains to determine that the sand from Iceland is different from the sand from Dockweiler Beach? List two additional tests you could perform on the sand.

_____

_____

13. The black sand from Iceland is composed of obsidian, a black volcanic glass. Zoom in to the maximum and look at the grains closely. What evidence can you see to suggest that this glass came from volcanic rocks?

_____

Sand #6: Olivine Sand from Toilet Bowl, Oahu, Hawaii

http://faculty.pasadena.edu/dndouglass/sand/SandPile/TBowl.htm

The Toilet Bowl is a natural pool in basaltic lava rock along Hanauma Bay that is connected to the ocean by an underwater channel. The water level in the pool rises and falls quickly by about 5 feet as water surges in and out from below. (See the online video http://www.youtube.com/watch?v=vso7OVi4Ioc.)

Sand #7: Olivine Sand from Green Sand Beach, Hawaii

http://faculty.pasadena.edu/dndouglass/sand/SandPile/GrnSand.htm

Sand #6 and Sand #7 are green sand, composed almost entirely of the mineral **olivine**.

14. Zoom in on each of the olivine sands. Using the sandstone roundness guides (Figure 3.9), describe the roundness of each of the sands. Are they well-rounded, rounded, subrounded, subangular, or angular?

Sand Specimen	Roundness
Toilet Bowl, Oahu, Hawaii	
Green Sand Beach, Hawaii	

15. Where do you think the olivine comes from? _____

    Why? Explain your reasoning. _____

16. Which of the two sands has the larger grains? _____

17. From your observation of grain size, predict which beach has the higher wave energy. _____

18. How does grain size relate to roundness? Are the larger grains rounder or less rounded? _____

Sand #8: Oolitic Sand from the Great Salt Lake, Utah

http://faculty.pasadena.edu/dndouglass/sand/SandPile/SltLake.htm

This is a lake sediment. Sand like this is also found in warm shallow seas in tropical regions. Look at this sand carefully under high magnification. These grains are called **oolites** (or ooids), and they are composed entirely of calcium carbonate.

19. Using the sandstone roundness guides (Figure 3.9), describe the roundness of the sand. Is it well-rounded, rounded, subrounded, subangular, or angular?

    _____

20. Using the sandstone sorting guide (Figure 3.8), describe the sorting of this sand. Is it very well-sorted, well-sorted, moderately well-sorted, poorly sorted, or very poorly sorted?

    _____

21. Interpret the energy level of the depositional environment of this sand.

    _____

22. Go online to research the origin of oolitic sand. Write a paragraph on the origin of oolites. Use the following references:

    http://geology.utah.gov/utahgeo/rockmineral/collecting/oolitic.htm

    http://www.nysm.nysed.gov/virtual/collections/splendor_in_stone/splendortour17.html

    http://books.google.com and search for:

    1. *Encyclopedia of Sediments and Sedimentary Rocks* oolites, page 503

    2. *Microfacies of carbonate rocks*, pages 142–156

    _____

    _____

    _____

Sand #9: Eolian Sands from Al Wasrah, Kuwait

http://faculty.pasadena.edu/dndouglass/sand/SandPile/AlWasra.htm

Eolian sands differ from other types of sands in that they have been deposited by the action of wind.

23. Using the sandstone roundness guides (Figure 3.9), describe the roundness of the sand. Is it well-rounded, rounded, subrounded, subangular, or angular?

    _____

24. Using the sandstone sorting guide (Figure 3.8), describe the sorting of this sand. Is it very well-sorted, well-sorted, moderately well-sorted, poorly sorted, or very poorly sorted?

    _____

25. Compare the eolian sand with the beach sands. List three differences.

    _____

    _____

    _____

QUANTITATIVE STUDY OF SAND: ANALYSIS OF HEAVY MINERAL SAND

This exercise involves examining an analysis of a heavy mineral sand sample from one of the Georgia barrier islands, doing some calculations, and drawing a pie diagram. For this exercise, you should use a computer with spreadsheet software, such as Excel. If a computer is not available, you may use a calculator, a drawing compass (a point on one end and a pencil on the other), a protractor (to measure and plot angles on the circle you will draw), and a ruler.

1. Examine the composition of the heavy mineral sand in the table below. The sample weight was 147 g, and the weight of each mineral is given in grams. Calculate the percentage of each mineral present in the sample. To check your work, add all the weights to confirm that the sample weighed 147 g. Then add the calculated percentages to confirm that the total equals 100%. Write your answers on the table below or attach a printout from your spreadsheet program. Attach a separate sheet with your pie diagram.

Mineral	Sample Weight (g)	% Total Weight
Ilmenite	74.59	
Zircon	32.74	
Staurolite	9.94	
Epidote	8.57	
Rutile	7.97	
Leucoxene	2.63	
Monazite	2.54	
Kyanite/sillimanite	2.25	
Garnet	2.23	
Quartz	1.21	
Hornblende	1.18	
Tourmaline	1.00	
Corundum	0.15	
Other	0.0	
TOTAL	147 g	100%

Heavy mineral data from Bishop, G.A. and, Marsh, N.B. Activity 3. The Geology of Georgia, Virtual Field Trip to Georgia Coast, 1997.

2. Draw a pie diagram to illustrate the percentages of each of the heavy mineral components. This is easily done by computer using Excel. If you do not have access to a computer, you may draw your pie diagram by hand, following these instructions.

   - Using a drawing compass, draw a circle with a radius of about 2 inches. Do not draw the circle freehand. You must use a drawing compass.
   - Using a calculator, determine the relative proportions of each mineral from the data provided in terms of degrees of arc. *Example*: Zircon = 22.27%, therefore 360 degrees × 0.2227 = 80.172 degrees, which can be plotted to the nearest degree—in this case, 80 degrees.
   - Use a protractor to measure the angle in degrees, and draw a line using a ruler.
   - Plot the number of degrees of arc in the pie to represent each mineral.
   - Each mineral should be identified in a legend and colored on the pie chart.

OPTIONAL ACTIVITY

Start a sand collection. You might be able to find sand at a nearby stream, river, lake or beach. You can also collect sand when you travel. There are several websites related to sand collecting.

http://sandcollectors.org/
http://www.microscope-microscope.org/applications/sand/sand-links.htm
http://www.scienceart.nl/Frames/HOMEpage.htm

Stream cobbles, Henson Creek, Prince George's County, Maryland.

Pamela Gore

Sedimentary Rocks

Sedimentary rocks are important because they contain the historical record of ancient environments and life on Earth. Throughout this course we will be studying sedimentary rocks, the fossils they contain, and the history that they record.

Sediment is loose particulate material, which can form in several ways. Sediment may be derived from the weathering and erosion of preexisting rock (called **terrigenous**, or **detrital**, or **clastic** sediment). Sediment also may be formed from chemical, biochemical, or biological materials, such as minerals formed by the evaporation of seawater, seashells, or plant remains. **Sedimentary rocks** are formed when sediment is compacted and cemented together. Approximately 75% of the rocks exposed at Earth's surface are sedimentary rocks.

Sediment accumulates in subaqueous environments, such as lakes, rivers, bays, deltas, beaches, and ocean basins. Sediment also may be deposited in other types of environments, such as deserts or glaciated areas. The characteristics of the sediment (grain size, shape, sorting, and composition), and the sedimentary structures are clues to the environment in which the sediment was deposited. In general, it takes more energy (greater water velocity) to transport larger grains.

CLASSIFICATION OF SEDIMENTARY ROCKS

Sedimentary rocks are grouped according to their origin:

- **Terrigenous sedimentary rocks** (also called clastic or detrital sedimentary rocks) form from fragments of preexisting rocks.
- **Chemical and biochemical sedimentary rocks** form as chemical precipitates or from the shells of organisms.
- **Organic sedimentary rocks** are composed of organic matter or carbon.

TERRIGENOUS (CLASTIC OR DETRITAL) SEDIMENTARY ROCKS

Terrigenous sedimentary rocks are derived from preexisting rocks. They are composed of rock fragments and mineral grains that have been weathered, eroded, transported, deposited, and cemented together to form a sedimentary rock. They are sometimes referred to as **extrabasinal** because they are derived from rocks *outside of the basin* of deposition. The individual grains (or clasts) in these rocks are mechanically durable (able to withstand abrasion during transport) and chemically stable. Typical clasts are made of quartz, feldspar, muscovite, clay minerals, or rock fragments.

Texture

Most clastic sedimentary rocks have three textural components (Figure 4.1).

1. **Clasts** are larger pieces of sedimentary debris (gravel, sand, silt).
2. **Matrix** is fine-grained material surrounding clasts.
3. **Cement** is the chemical glue that holds the rocks together. The most common cements are calcite, iron oxide, and silica cement.

   a. **Calcite cement** makes any sedimentary rock fizz in hydrochloric acid.
   b. **Iron oxide cement** gives the rock a reddish brown, pink, red, or orange color.
   c. **Silica** (quartz) **cements** do not fizz in hydrochloric acid and are not reddish brown. They are best recognized by process of elimination.

Clast Size

Most terrigenous sedimentary rocks are classified by the size of the clasts of sediment they contain. The size ranges of sedimentary grains are shown in Tables 4.1 and 4.2. Table 4.1 gives a simplified version to memorize, and Table 4.2 gives a more complex version for reference.

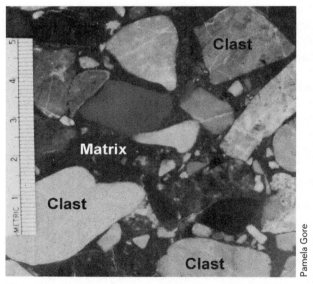

Figure 4.1 Clasts and matrix (labeled), and iron oxide cement (reddish brown color surrounding clasts). (Scale in millimeters.)

TABLE 4.1 SIMPLIFIED GRAIN SIZE CLASSIFICATION		
Particle Name		**Particle Diameter**
Gravel		> 2 mm
Sand		1/16 - 2 mm
Mud	Silt	1/256 - 1/16 mm
	Clay	<1/256 mm

TABLE 4.2   CLASSIFICATION OF GRAIN SIZE

Particle Name	Particle Diameter	Textural Term
Gravel		
Boulders	> 256 mm	Rudite
Cobbles	64–256 mm	
Pebbles	4–64 mm	
Granules	2–4 mm	
Sand		
Very coarse sand	1–2 mm	
Coarse sand	0.5–1 mm	
Medium sand	0.25–0.5 mm	Arenite
Fine sand	0.125–0.25 mm	
Very fine sand	0.0625–0.125 mm	
Mud		
Silt	0.004–0.0625 mm (1/256 to 1/16 mm)	Argillite
Clay	< 0.004 mm (<1/256 mm)	

Textural Terms

Rudites: Sedimentary rocks with gravel-sized clasts are sometimes referred to as rudites. *Rudite* means "gravel."

Arenites: Arenaceous sedimentary rocks or arenites are those with sand-sized grains. *Arenite* means "sand." The word is derived from the sand that covered the floor of the Roman arenas where the gladiators fought.

Argillites: Argillaceous sedimentary rocks or argillites are those with mud. (Mud is defined as a mixture of silt and clay.) *Argillite* means "mud."

Clast Shape

Shape of clasts is important in naming sedimentary rocks with gravel-sized clasts. Gravel may be *rounded* or *angular* (based on the sharpness of the corners of the clasts). Gravel rapidly becomes rounded in the first few miles of transport.

Sorting

Sorting refers to the distribution of grain sizes in a rock. If all of the grains are the same size, the rock is *well-sorted*. If there is a mixture of grain sizes, such as sand and clay, or gravel and sand, the rock is *poorly sorted*.

Rocks with Gravel-Sized Clasts

Conglomerate (Figure 4.2) and **breccia** (Figure 4.3) contain gravel-sized clasts surrounded by finer-grained matrix. Conglomerate has *rounded* clasts. If the particles are *angular*, the rock is a breccia. In a conglomerate, the larger clasts are generally more rounded than the smaller clasts.

A. Conglomerate.

B. Conglomerate with rounded quartz clasts in a sandy matrix. (Scale in centimeters.)

Figure 4.2 Conglomerate.

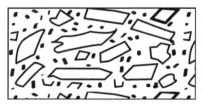

A. Breccia.

B. Breccia with angular clasts of dolostone in a sandy matrix. (Scale in centimeters.)

Figure 4.3 Breccia.

Rocks with Sand-Sized Clasts

Sandstones contain sand-sized clasts. Sand grains may be either rounded or angular, and they are generally more or less the same size (*well-sorted*). The sand grains are held together by cement, which may be *silica* (quartz), *calcite*, or *iron oxide*. Sandstones are classified according to the composition of the sand grains into three main groups (Figure 4.4):

Quartz sandstone or **quartz arenite** is composed mainly of *quartz* sand grains.

Arkose is composed mainly of pink or white *feldspar* grains, with quartz and generally some muscovite mica or sand-sized rock fragments.

Litharenite (meaning "rock-sand") or **lithic sandstone** or **graywacke** is predominantly composed of dark sand-sized **rock fragments**, with some mica, quartz, and feldspar grains in a clay-rich matrix. A *wacke* is defined as a "dirty" (or muddy) sand. The term *graywacke* is best used loosely; there is no strict definition of the term with which all geologists agree. A litharenite is more strictly defined as a rock primarily composed of sand-sized rock fragments.

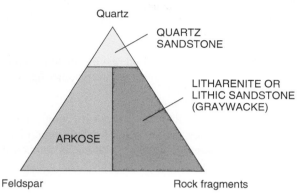

A. Simplified sandstone classification diagram.

B. Quartz sandstone or quartz arenite. (Scale in centimeters.)

C. Arkose with pink feldspar grains. (Scale in centimeters.)

D. Litharenite or lithic sandstone, sometimes called graywacke. (Scale in centimeters.)

Figure 4.4 Types of sandstone.

Rocks with Silt-Sized Grains

Siltstone is intermediate in texture between sandstone and shale (Figure 4.5). The grains are difficult to see with the naked eye because they are so small, but the rock has a distinct *gritty feel* to the fingernails.

Figure 4.5 Siltstone. (Scale in millimeters.)

Clay-Dominated Rocks

Shale and **claystone** are fine-grained sedimentary rocks composed of tiny (<1/256 mm) grains of clay minerals, mica, and quartz. The individual grains are too small to see with the naked eye or a hand lens, and the rock feels smooth to the touch, not gritty.

Shale and claystone differ in the way that they break (Figure 4.6). Shale is *fissile*. This means that it splits readily into thin, flat layers. Claystone, on the other hand, is massive, and it breaks irregularly.

- The color of shale or claystone can reveal something about its composition. **Black shales** contain organic matter. They are sometimes called bituminous shales. **Red shales** contain iron oxide. **Kaolin**, a white claystone, is composed of the mineral kaolinite. It lacks organic matter and iron oxide.

Mud

Mud is defined as a mixture of silt and clay. Rocks with both silt and clay are referred to as **mudstones** or **mudshales**, depending on whether or not they are fissile.

A. Shale is *fissile*. It splits into flat layers. (Scale in centimeters.)

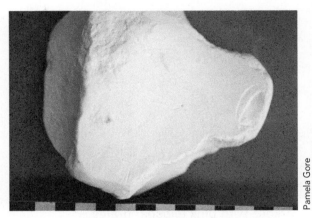

B. Claystone is massive. White claystone is called *kaolin*. Sandersville, Georgia. (Scale in centimeters.)

Figure 4.6 Clay-dominated rocks: shale and claystone.

CHEMICAL AND BIOCHEMICAL SEDIMENTARY ROCKS

Chemical, biochemical, and organic sedimentary rocks are sometimes referred to as *intrabasinal* because they form within the basin of deposition, rather than being transported into it. They include **chemical precipitates** (such as travertine, rock salt, and gypsum), as well as the accumulated remains of organisms that lived within the basin (such as limestones composed of fossil shells). We divide chemical and biochemical sedimentary rocks into four groups: evaporites, siliceous rocks, ironstones, and carbonate rocks.

Evaporites

Evaporites are *chemical precipitates*, which form by precipitation of dissolved minerals from water during evaporation. There are numerous evaporites, but we concentrate on three: travertine, rock gypsum, and rock salt.

Travertine (Calcite, CaCO₃)

Travertine forms by evaporation of cave, spring, or river waters. It consists of inter-grown calcite crystals, and it fizzes in hydrochloric acid. Travertine is a dense, crystalline rock with tan and white bands. It is especially common in limestone caverns, where it forms flowstone and dripstone, including *stalactites* and *stalagmites*, recognized by their cylindrical shape and internal tree ring-like appearance. Travertine that forms around springs is a more porous, light-colored rock that is sometimes called tufa (Figure 4.7). Because travertine is composed of calcite, it is also mentioned later with the carbonate rocks.

A. Laminated travertine in a stalactite. (Scale in centimeters.)

B. Travertine with vague layering and large pore spaces from a hot spring deposit. Travertine is commonly used as tile and decorative stone. (Pencil for scale.)

Figure 4.7   Travertine.

Rock Gypsum (Gypsum, CaSO₄•2H₂O)

Rock gypsum is softer than your fingernail (you can scratch it). It can be granular, can be glassy with obvious cleavage, or can have a fibrous, satiny luster. It ranges from colorless to white to pink (Figure 4.8). Gypsum forms by precipitation from seawater after about 80% of the water has evaporated. Gypsum is altered to **anhydrite** (CaSO₄) by removal of water, which is generally caused by burial to depths greater than several hundred meters. For purposes of this lab, you do not need to distinguish between gypsum and anhydrite.

15 cm

Figure 4.8   Laminated gypsum, Castile Formation (Permian), Carlsbad, New Mexico. (Scale in centimeters.)

Rock Salt (Halite, NaCl)

Rock salt is a glassy, crystalline rock that is easily recognized by its salty taste. It ranges from colorless, to white, gray, pink, or orange, due to impurities (Figure 4.9). Halite forms by precipitation from sea water after about 90% of the water has evaporated.

Figure 4.9   Rock salt. (Scale in millimeters.)

Siliceous Sedimentary Rocks

Siliceous rocks are dominated by silica (SiO_2), which precipitates from solution within the basin of deposition. (They do not include quartz sandstones, which are extrabasinal in origin.) The most common siliceous sedimentary rocks are diatomite and chert.

Diatomite

Diatomite (diatomaceous earth) is a soft, white, powdery rock of low density, composed of the siliceous (silica) skeletons of microscopic algae called **diatoms** (Figure 4.10). Diatomite can be distinguished from chalk because it does not react with hydrochloric acid. It can be distinguished from kaolinite by its low density. Diatomite floats on water.

Chert (Microcrystalline Quartz, SiO₂)

Chert is a very fine grained silica sediment of chemical or biochemical origin. Some chert contains siliceous skeletons of microorganisms known as *radiolarians*, which can be seen in thin section (Figure 4.11). Other chert forms through the replacement of limestone, often preserving carbonate textures such as oolites, although the rock has been completely altered to silica.

Figure 4.10   Diatomite. (Scale in millimeters.)

Figure 4.11   Chert. (Scale in centimeters.)

Chert can be recognized by its extremely fine grain size, smooth feel, and hardness (it scratches glass). Chert may be black, white, tan, gray, or greenish gray. A red variety is called jasper. Flint is sometimes used as a synonym for chert, but the term is used loosely and is best reserved for artifacts, such as arrowheads.

Sedimentary Ironstones

Some sedimentary rocks are dominated by iron-bearing minerals such as hematite. Common examples of sedimentary ironstones include **banded iron formations**, **oolitic hematite or oolitic ironstone**, and **iron oxide concretions**, typically in sand (Figure 4.12). Banded iron formations consist of layers of red chert (jasper) containing oxidized iron, alternating with gray, unoxidized iron (hematite or magnetite). Banded iron formations are Precambrian (Paleoproterozoic and Archean) in age, 2.6 to 1.8 billion years old.

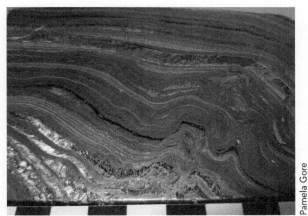

A. Banded iron formation. This specimen also has a few layers of yellow tiger's eye or silicified asbestos. (Scale in centimeters.)

B. Oolitic ironstone, Silurian Red Mountain Formation, Birmingham, Alabama. (Scale in centimeters.)

C. Iron oxide concretion, Lumpkin, Georgia. (Scale in centimeters.)

D. Road cut through oolitic ironstone of the Silurian Red Mountain Formation, Birmingham, Alabama. (Geology students for scale.)

Figure 4.12 Sedimentary ironstones *(Continued).*

E. Dark, irregular layer of iron oxide cemented sand and concretions overlying white Providence Sand (Cretaceous). Iron layer is 10 to 15 cm thick. (Lumpkin, Georgia.)

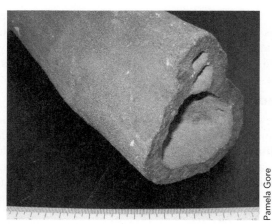

F. Iron oxide concretions are commonly hollow and irregularly shaped. (Scale in millimeters.)

G. Iron oxide concretion, Hyattsville, Maryland. (Scale in centimeters.)

Figure 4.12 Sedimentary ironstones.

Carbonate Rocks

Limestone and **dolostone** are called carbonate rocks because they are composed of carbonate minerals, which have a CO_3^{2-} group in their chemical formula. The carbonate minerals include calcite and aragonite (both $CaCO_3$, calcium carbonate) and dolomite ($CaMg(CO_3)_2$, calcium-magnesium carbonate.

Note that calcite and aragonite are **polymorphs** of calcium carbonate. Calcite is the *stable* form, and aragonite is *metastable*. Aragonite alters to calcite over a long time. (It is not necessary to distinguish calcite from aragonite in the lab.) Rocks that contain abundant calcium carbonate are often referred to as *calcareous* rocks (derived from the word *calcite*).

Limestone is primarily composed of the minerals calcite and aragonite. Limestones are generally gray, but may be tan, pink, white, black, or other colors.

Dolostone is primarily composed of the mineral **dolomite**. Weathered surfaces of dolostones are commonly yellowish or brownish gray because of the presence of small amounts of iron associated with the magnesium in dolomite. Dolomite results from the addition of magnesium to calcium carbonate sediments after deposition, in arid climates. Magnesium chemically replaces some of the calcium in calcite to turn a limestone into a dolostone.

Carbonate rocks and minerals are easy to identify because they react with hydrochloric acid. Limestone effervesces (fizzes) readily in hydrochloric acid. Dolostone fizzes weakly in hydrochloric acid, but only after it has been powdered.

Carbonate rocks most commonly form in *warm shallow seas* in places such as the Florida Keys and the Bahamas (Figure 4.13).

Figure 4.13 Carbonate tidal flats in the Florida Keys.

Calcareous algae are plants with calcium carbonate skeletons. They live in *warm shallow seas* in areas such as Florida Bay (Figure 4.14). When calcareous algae die, the organic tissues decompose, releasing fine particles of calcium carbonate (aragonite) that accumulate to form lime mud. Lime mud undergoes recrystallization to form fine-grained limestone or **micrite**, which is an abbreviation for *micr*ocrystalline calc*ite*.

Figure 4.14 Calcareous algae from Florida Bay.
Note the large holdfast (root ball) at the bottom. *Penicillus,* the "shaving brush algae," is on the *left. Halimeda* is on the *right.* (Scale in centimeters.)

Textures of Carbonate Rocks

Terrigenous rocks are composed of clasts, matrix, and cement. Carbonate rocks have similar textural components: allochems, matrix, and spar.

Allochems are particles in carbonate rocks, a term that is analogous to clasts in terrigenous rocks. Types of allochems are intraclasts, oolites, fossils, and pellets.

Matrix is microcrystalline calcite or lime mud.

Spar is calcite cement.

The textures of carbonate rocks are best studied in *thin sections* (thin slices of rock mounted on glass microscope slides.

Classification of Carbonate Rocks

There are numerous types of carbonate rocks, and they can be classified by their textures and the allochems that they contain. There are several different classification schemes for carbonate rocks. One of the simpler classifications is based on texture, and uses the following terms:

Calcirudite: Limestone dominated by *gravel*-sized particles

Calcarenite: Limestone dominated by *sand*-sized particles

Calcisiltite: Limestone dominated by *silt*-sized particles

Calcilutite: Limestone dominated by *mud-* or *clay*-sized particles

This classification is fairly general, and it does not specify anything about the types of allochems present.

Types of Carbonate Rocks

Micrite is fine-grained limestone (Figure 4.15). The color of micrite ranges from gray to tan or other colors. Micrite is an abbreviation for *micro*crystalline calc*ite. Microcrystalline* refers to the texture, which consists of clay-sized particles of lime mud. Basically, this is a rock that is all matrix with no allochems (larger particles) or spar (calcite cement). Calcilutite is another name for a limestone dominated by lime mud.

Fossiliferous limestone, sometimes called *skeletal limestone*, contains *fossils*, or the remains of ancient plants, animals, or algae. Many organisms have calcareous shells or skeletons, and their remains can accumulate in lime mud to form fossiliferous limestone (Figure 4.16).

Coquina is a type of fossiliferous limestone made up of fossil shells with little or no matrix (Figure 4.17). It is porous and light-colored, and the shells are commonly broken, abraded, and fairly well-sorted. The shells are gravel sized (>2 mm), and coquina is a calcirudite.

A. Gray micrite with brown dolostone layers. (Scale in centimeters.)

B. Outcrop of micrite. (Geology students for scale.) What type of fold is this?

Figure 4.15 Micrite

A. Fossiliferous limestone (Ordovician), northwestern Georgia. (Scale in centimeters.)

Figure 4.16 Fossiliferous limestone.

B. Fossiliferous limestone with brachiopod fossils, western Maryland. (Scale in centimeters.)

Figure 4.17 Coquina, St. Augustine, Florida. (Scale in centimeters.)

Chalk is a type of fossiliferous limestone made up entirely of microscopic shells (Figure 4.18). These tiny shells are called **coccoliths** and are the remains of planktonic marine algae called **coccolithophores**. Coccoliths are typically too small to see using an ordinary light microscope (although they may be seen using an oil immersion lens). They are generally viewed using an electron microscope. The texture of chalk is similar to that of micrite or calcilutite, but chalk is white, less dense, and less compact than micrite. Chalk may be distinguished from other white fine-grained sedimentary rocks (such as kaolinite or diatomite) because it fizzes readily in hydrochloric acid.

Figure 4.18 Chalk is white and fine-grained. It can be distinguished from diatomite and kaolinite because chalk fizzes in hydrochloric acid. Specimen is about 10 cm wide.

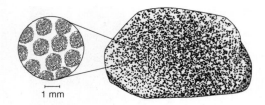

A. Oolitic limestone.
(Scale in centimeters).

B. Oolitic limestone with black oolites.
(Scale in millimeters).

Figure 4.19

Oolitic limestone is composed of **oolites** (Figure 4.19). Oolites are tiny concentric spheres of calcium carbonate that range between 0.1 and 2.0 mm in diameter. On a cut or broken surface they look circular, and internal concentric laminations may be seen with a hand lens or microscope. Oolites are *not* fossils! They form by the precipitation of aragonite under certain conditions in warm shallow seas (or salt lakes) under the influence of blue-green algae. Because oolites are less than 2 mm in diameter (sand sized), oolitic limestone is *calcarenite*. Structures that resemble oolites and that are larger than 2 mm in diameter are called *pisolites*.

Intraclastic limestone is a limestone conglomerate or breccia with a lime mud matrix (Figure 4.20). **Intraclasts** are flat, gravel-sized chips of limestone that form when carbonate tidal flats that are covered by lime mud dry up, experience cracking, and break into flat, gravel-sized chips. These chips of lime mud are redistributed by the tides and currents, and they accumulate to form intraclastic limestone. Intraclasts may be internally layered, reflecting the layering in the tidal flat sediments. Intraclastic limestone is *calcirudite*.

Pelleted limestone is dominated by **pellets**, which are small (<1 mm) aggregates of microcrystalline calcite, many of which are fecal in origin (excrement of marine invertebrates, such as clams and worms). Unlike oolites, they have no internal structure. Pellets are so small that they generally cannot be seen in hand specimens, but they can be seen in thin sections using a microscope.

Crystalline limestone generally consists of a coarse mosaic of intergrown calcite crystals, resulting from the post-depositional alteration of some other type of limestone. Allochems might or might not be visible.

Figure 4.20 Intraclastic limestone. (Scale in centimeters.)

Travertine is a dense crystalline limestone with tan and white color bands that forms in caves or hot springs (Figure 4.21). It consists of a mosaic of intergrown calcite crystals. In lab, stalactites and stalagmites can be recognized by their cylindrical shape and internal tree ring-like appearance. Texturally, travertine is essentially a carbonate rock made up entirely of spar (calcite cement).

Figure 4.21 Travertine in a stalactite. (Scale in millimeters.)

Figure 4.22 Interlayered limestone (gray) and dolostone (tan), Conococheague Limestone (Cambrian), Clear Spring, Maryland. (Scale in centimeters.)

Dolostone is made up of the mineral dolomite, a calcium-magnesium carbonate (Figure 4.22). Most dolostones form by the chemical replacement of calcium carbonate through the action of magnesium-rich fluids. Magnesium chemically replaces some of the calcium in calcite to turn a limestone into a dolostone. A dolostone might retain some of the texture of the original limestone, but it is typically dense and compact, with a fine-grained texture. Dolomite fizzes weakly in hydrochloric acid, and only after the rock is scratched or powdered. Dolomite is harder than calcite.

ORGANIC SEDIMENTARY ROCKS: COAL

Coals are organic sedimentary rocks composed primarily of carbon-rich organic material derived from the remains of plants. Unlike the other sedimentary rocks, coals are not mainly composed of minerals. Minerals are by definition *in*organic, and coal is mostly made of *organic* matter.

Peat is sediment composed of plant fragments. Coal is its lithified equivalent. The plant fossils in coal generally indicate deposition in fresh-water swamps. Peat is transformed by burial pressure and temperature to **lignite** (a soft, black or brownish, coal-like material). Lignite alters to sooty **bituminous coal** with greater depth and duration of burial, and higher temperatures. With increasing metamorphism (such as proximity to an igneous intrusion, or increasing temperatures and pressures associated with burial), bituminous coal alters to **anthracite coal**, a hard, shiny coal (Chart 4.1). Geologists consider anthracite to be, in fact, a metamorphic rock.

Chart 4.1 Transformation of Peat to Anthracite Coal

Peat → Lignite → Bituminous coal → Anthracite coal (metamorphic)

Classification of Sedimentary Rocks

Sedimentary rocks can be classified by whether or not they have visible grains. Sedimentary rocks without visible sedimentary grains can be classified by whether or not they fizz in hydrochloric acid. Table 4.3 lists rocks with visible grains. Table 4.4 lists rocks that do not have visible grains.

Instructions: Examine the sedimentary rock and make a series of decisions to determine the rock name. First, determine whether the grains in the rock are clearly visible. If so, use Table 4.3. Next, determine whether the clasts or allochems are larger or smaller than 2 mm. Then, read the descriptions to identify the sedimentary rock.

If grains are *not* readily visible use Table 4.4. Test the rock with hydrochloric acid to see if it fizzes readily, if it only fizzes after it has been scratched and powdered, or if it does not react at all with hydrochloric acid. Then, read the descriptions to identify the sedimentary rock.

TABLE 4.3 SEDIMENTARY ROCKS WITH VISIBLE GRAINS

Grain Size		Description	Sedimentary Rock Name
Grains visible	Clasts or allochems larger than 2 mm	Grains are all shell fragments; no mud; fizzes in hydrochloric acid	**Coquina**
		Clasts and matrix are fine grained; clasts are limestone and may be flat and laminated; fizzes in hydrochloric acid	**Intraclastic limestone**
		Matrix color variable; multiple clast lithologies; clasts differ from matrix in color or composition	**Breccia** (angular clasts)
			Conglomerate (rounded clasts)
	Clasts or allochems smaller than 2 mm	White or colorless grains, mostly quartz	**Quartz sandstone**
		Contains pink, gray, or white feldspar (look for cleavage); feldspar grains may be weathered to white kaolinite	**Arkose**
		Contains rock fragment grains, mostly dark green or gray grains (such as basalt or shale fragments)	**Litharenite** or **lithic sandstone** or **graywacke**
		Round grains with concentric laminations, fizzes in acid	**Oolitic limestone**
		Dark red to brown, red-brown streak, may contain replaced oolites or fossils, may fizz in hydrochloric acid, may be dense and heavy	**Oolitic ironstone** or **fossiliferous ironstone** or **iron oxide concretions**

TABLE 4.4 SEDIMENTARY ROCKS GRAINS THAT ARE NOT READILY VISIBLE

Grain Size	Reaction to Acid	Description	Sedimentary Rock Name
Grains not visible	Fizzes in HCl acid	White, soft, and powdery	**Chalk**
		Gray, black, brown or tan; compact; dense; very fine grained (clay-sized)	**Micrite** or **calcilutite**
		Fossils in lime mud matrix	**Fossiliferous limestone**
		Coarse crystalline mosaic; brown and white color bands; may be cylindrical (stalactite or stalagmite)	**Travertine**
		Fine to coarse crystalline mosaic; compact, dense, massive	**Crystalline limestone**
	Fizzes in acid when scratched and powdered	Gray or black, weathers yellowish gray to brown; compact, dense, massive; dolomite	**Dolostone**

(Continued)

TABLE 4.4 *(Continued)*

Grain Size	Reaction to Acid	Description	Sedimentary Rock Name
Grains not visible	Does not fizz in acid	Fissile (breaks into thin layers); may be softer than fingernail; clay-sized texture; commonly gray, black, brown or red	**Shale**
		Feels gritty to the fingernails; commonly gray, black, brown, or red	**Siltstone**
		Salty taste, may feel slippery; often clear and transparent; cleavage	**Rock salt**
		Softer than fingernail; white, pink, clear; may be fibrous, fine-grained, or crystalline	**Rock gypsum**
		Hard - scratches glass; opaque; color variable; smooth feel; may have conchoidal fracture	**Chert**
		White; looks like chalk but does not fizz; very low density (may float)	**Diatomite**
		White; looks like chalk but does not fizz; dense (does not float); may stick to moistened finger	**Kaolinite**
		Black, bright and shiny (may almost look metallic in luster), compact, low density	**Anthracite coal**
		Black, may leave sooty marks on fingers or paper, may have layers	**Bituminous coal**
		Brown to black, crumbly, very soft; porous	**Lignite**
		Brown, porous, soft; pieces of plants may be visible	**Peat**

Summary

Here is a list of the major rock types mentioned in this lab.

Terrigenous (or Clastic, or Detrital) Sedimentary Rocks

1. Conglomerate
2. Breccia
3. Sandstone
   - Quartz sandstone or quartz arenite
   - Arkose
   - Litharenite or lithic sandstone or graywacke
4. Siltstone
5. Shale
6. Claystone
7. Mudstone
8. Mudshale

Chemical and Biochemical Sedimentary Rocks

EVAPORITES

1. Travertine
2. Rock gypsum
3. Rock salt

SILICEOUS SEDIMENTARY ROCKS

1. Diatomite
2. Chert

SEDIMENTARY IRONSTONES

1. Banded iron formation
2. Oolitic ironstone
3. Iron oxide concretions

CARBONATE ROCKS: LIMESTONE AND DOLOSTONE

1. Micrite
2. Fossiliferous limestone
3. Coquina
4. Chalk
5. Oolitic limestone
6. Intraclastic limestone
7. Pelleted limestone
8. Crystalline limestone
9. Travertine
10. Dolostone

Organic Sedimentary Rocks

COAL

1. Peat
2. Lignite
3. Bituminous coal
4. Anthracite coal (metamorphic)

Pamela Gore

Beach sand and shells, St. Augustine, Florida. Shells average about 1 cm in diameter.

Sedimentary Rocks Exercises

PRE-LAB EXERCISES

Answer these questions *before* looking at the sedimentary rock specimens to be sure that you have mastered the material in the lab.

1. Which of the following are *terrigenous* sedimentary rocks? Draw a circle around them.

Black shale	Arkose	Intraclastic limestone
Diatomite	Rock gypsum	Quartz conglomerate
Oolitic limestone	Red sandstone	Lignite
Graywacke	Chert	Bituminous coal

2. There are three white, fine-grained sedimentary rocks. What distinguishing characteristics might you use to tell them apart?

   Chalk _____

   Kaolin _____

   Diatomite _____

3. What is the main difference between shale and claystone?

   _____

4. For each of the rock types in the table below, give the name of the dominant mineral, the chemical formula of that mineral, and tell whether or not the mineral (or rock) will effervesce (fizz) in hydrochloric acid.

Rock Type	Mineral Name	Chemical Formula	Reaction to HCl
Limestone			
Dolostone			
Rock gypsum			
Rock salt			
Chert			
Travertine			

5. Classify the following terrigenous rocks according to whether they are:
(a) rudites, (b) arenites, or (c) argillites. Put the letter in the blank.

_____ Lithic sandstone

_____ Conglomerate with quartz clasts

_____ Arkose

_____ Gray silty shale

_____ Breccia with iron oxide cement

_____ Kaolinite

6. Classify the following carbonate rocks according to whether they are
(a) calcirudites, (b) calcarenites, or (c) calcilutites. Put the letter in the blank.

_____ Intraclastic limestone

_____ Chalk

_____ Oolitic limestone

_____ Coquina

_____ Micrite

7. Match the rock type with the sediment from which it is formed. Put the letter in the blank.

_____ micrite	a. rounded gravel
_____ shale	b. plant fragments
_____ graywacke	c. microscopic skeletons of silica
_____ chalk	d. feldspar grains
_____ diatomite	e. angular gravel
_____ lignite	f. sand-sized rock fragments
_____ coquina	g. clay
_____ arkose	h. lime mud
_____ conglomerate	i. broken shells
_____ breccia	j. coccoliths

8. List three allochems you might find in limestone.

_____

_____

_____

LAB EXERCISES

Sedimentary Rock Identification

Instructions

Examine the sedimentary rock samples provided by your instructor. Fill in the chart below, and use Table 4.3 and Table 4.4 in the section Classification of Sedimentary Rocks to help you identify the samples.

Instructions for Filling in the Sedimentary Rock Identification Table
Column 1: Grain Size (If Visible)

There are three ways to measure grain size:

- Measure the size of the grains in millimeters with a ruler. This works well for gravel-sized clasts, but it may be difficult for finer-grained samples.
- Compare the grain size to that of reference sediment textures on a grain-size comparator or sand gauge. This works well for distinguishing sandstone from siltstone.
- List whether the sample consists of gravel, sand, silt, or clay particles based on a simple visual inspection.
- If you cannot see grains, write "Grains not visible."

Column 2: Minerals or Grain Type

- Identify the *minerals* present (calcite, dolomite, gypsum, halite, quartz, feldspar, clay minerals, etc.)
- If *grains* (clasts or allochems) are present, what are they? (Possibilities include fossils, oolites, intraclasts, and mineral grains or rock fragments.)

Column 3: Fizz in HCl?

- Use dilute hydrochloric acid to test the sample to determine whether calcium carbonate (calcite or aragonite) or dolomite is present. Calcite and aragonite readily fizz in acid. Dolomite fizzes only if scratched or powdered.
- If it fizzes readily, write "Yes."
- If it fizzes only when scratched and powdered, write "Only if powdered."
- If it does not fizz, write "No."

Column 4: Describe

- Use this column to describe the sample. For example, the color, any layers, or any other distinguishing features.

Column 5: Sedimentary Rock Name

- Identify the rock. Use Table 4.3 and Table 4.4 in the section Classification of Sedimentary Rocks.
- If you are in the lab, and if there are examples of "known" samples, you may also use them for comparison.

	Grain Size (if visible)	Minerals or Grain Type	Fizz in HCl?	Describe	Sedimentary Rock Name
1					
2					
3					
4					
5					

	Grain Size (if visible)	Minerals or Grain Type	Fizz in HCl?	Describe	Sedimentary Rock Name
6					
7					
8					
9					
10					
11					
12					
13					
14					
15					
16					
17					
18					
19					
20					
21					
22					
23					
24					
25					

OPTIONAL ACTIVITY

1. Examine samples of sand using a hand lens or a microscope, and answer the following questions. (Some questions might have more than one answer.)

_____ a. Which sand is better sorted?

_____ b. Which sand contains skeletons of microscopic marine animals?

_____ c. Which sample contains sand grains derived from rock weathering?

_____ d. Which sample might have been collected from a river?

_____ e. Which sample might have been collected from a warm, shallow sea?

2. Using a microscope, examine the thin sections provided by your instructor. Put the number of the thin section in the correct blank.

_____ a. Oolitic limestone

_____ b. Fossiliferous limestone

_____ c. Quartz sandstone

_____ d. Arkose

_____ e. Shale

_____ f. Siltstone

_____ g. Lithic sandstone (graywacke)

_____ h. Pelleted limestone

_____ i. Micrite or chalk

_____ j. Dolostone

Pamela Gore

Pedestal rock, mushroom rock, or hoodoo formed in quartz sandstone and conglomerate. Lookout Mountain, Georgia.

Sedimentary Structures

In this lab you will learn to recognize and identify sedimentary structures. Primary sedimentary structures are those that form during (or shortly after) the sediment is deposited. Some primary sedimentary structures are produced by water or wind that moves the sediment. Other primary sedimentary structures, such as footprints, worm trails, or mudcracks, form after deposition. Primary sedimentary structures can provide information about the environmental conditions under which the sediment was deposited because certain structures form in quiet water under low-energy conditions, whereas others form in moving water or high-energy conditions. Some form along the shore, some form in the deep sea, and others form in the open air. Primary sedimentary structures are important clues for the geologist attempting to interpret Earth history.

Sedimentary structures can be grouped into two basic types. **Inorganic sedimentary structures** are those formed by physical processes, like moving wind or water, or drying. **Organic** or **biogenic sedimentary structures** are those formed through the activities of animals or plants.

INORGANIC SEDIMENTARY STRUCTURES

Bedforms and Surface Markings

Bedforms and surface markings are sedimentary structures that form on the surface of a bed of sediment. At the time of formation, the surface of a bed is equivalent to the sea floor or the bottom of a lake or river. In a sequence of sedimentary rocks, bedforms and surface markings are found on the bedding planes. **Bedforms** are produced as water or wind moves across the sediment surface, and they include various types of **ripples**. Surface markings include **mudcracks** and **raindrop prints**.

Ripples are undulations of the sediment surface produced as wind or water moves across sand. Three types of ripples are covered in this lab: asymmetrical ripples, symmetrical ripples, and interference ripples. As ripples move under the influence of moving water or wind, **cross stratification** (an internal bedding structure with inclined layers) develops (see below). Ripples tend to form in *sand-sized sediment*.

Asymmetrical Ripples

Asymmetrical ripples form in *unidirectional currents* (such as in streams or rivers) (Figure 5.1). Crests of asymmetrical ripples may be straight, sinuous (curvy), or lingoid (lobelike or tongue-shaped), depending on water velocity. These ripples are asymmetrical in cross-sectional profile, with a gentle slope on the upstream side and a steep slope on the downstream side. Because of this unique geometry, asymmetrical ripples in the rock record may be used to determine ancient current directions or paleocurrent directions. Note that **cross stratification** (an internal bedding structure with inclined layers) forms as asymmetrical ripples slowly migrate downstream, and it can also be used as a paleocurrent indicator.

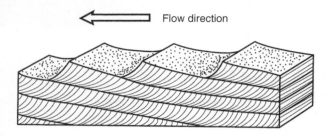

Flow direction

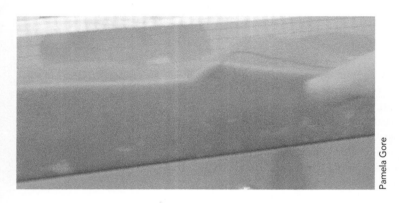

A. Asymmetrical ripples and cross stratification. From Gore, P. and Witherspoon, W. 2013, Roadside Geology of Georgia, Mountain Press Publishing Company, Missoula, MT.

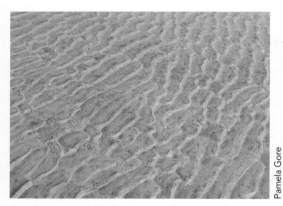

B. Asymmetrical ripple in sand in a flume tank. Water flow is from right to left. The red line marks the former position of the ripple crest. The ripple is migrating to the left, downstream. (Finger for scale.)

C. Asymmetrical current ripples, Georgia coast. Current flow is from left to right. (Pencil for scale in photo on *right.*)

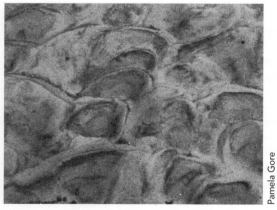

D. Asymmetrical lingoid ripples (tongue-like ripples). Jekyll Island, Georgia. Penny for scale.)

E. Asymmetrical lingoid ripples in sandstone (Paleozoic), northwestern Georgia. (Field of view is about 50 cm wide.)

Figure 5.1   Asymmetrical ripples.

Symmetrical Ripples

Symmetrical ripples are produced by *waves* or oscillating water (water moving back and forth) (Figure 5.2). Crests of symmetrical ripples tend to be relatively straight, but they can bifurcate (or fork). These ripples are symmetrical in cross-sectional profile. Cross stratification also forms beneath symmetrical ripples; however, the layers are curved and inclined in both directions as a result of the back-and-forth wave motion.

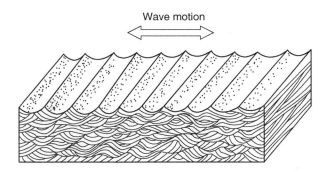

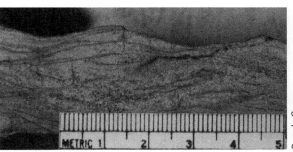

A. Symmetrical wave ripples and wave ripple cross stratification.
From Gore, P. and Witherspoon, W. 2013, Roadside Geology of Georgia, Mountain Press Publishing Company, Missoula, MT.

B. Symmetrical wave ripples, Georgia coast. (Pencil for scale.)

C. Side view (or cross-sectional view) of symmetrical wave ripples and wave-ripple cross stratification in sandstone (Triassic), Culpeper Basin, Virginia. (Scale in millimeters.)

D. Symmetrical wave ripples in red sandstone (Triassic), Culpeper Basin, Manassas, Virginia. (Specimen is about 30 cm across.)

E. Side view (or cross-sectional view) of symmetrical wave ripples in the specimen at left. (Width of view is about 10 cm.)

Figure 5.2 Symmetrical wave ripples.

Interference Ripples

Interactions between waves and currents can produce a more-complex pattern of **interference ripples** (Figure 5.3). This situation can occur along a coast where there is a series of sand bars attached to the coast at one end (called **a ridge and runnel system**) that is acted upon by both waves and tidal currents. At high tide, waves wash back and forth over the sand bars (ridges) and water-filled low areas between them (runnels), forming symmetrical wave ripples. When the tide goes out, currents form in the runnels as the water drains seaward in the low areas between the sand bars. Asymmetrical current ripples begin to form in the runnels, on top of the symmetrical wave ripples. Interference ripples contain remnants of wave ripples in one direction, with current ripples superimposed roughly perpendicular to them.

Mudcracks

Mudcracks are a polygonal pattern of cracks produced on the surface of mud as it dries (Figure 5.4).

Raindrop Prints

Raindrop prints are circular pits on the sediment surface produced by the impact of raindrops on soft mud (Figure 5.5).

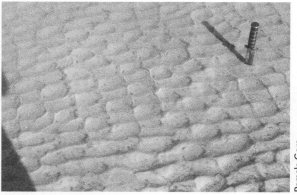

A. Ridge and runnel system at the south end of Jekyll Island, Georgia. Wave ripples are present on the sand bar, and current ripples form within the runnel (water-filled low area between sand bars) as the tide falls.

B. Interference ripples produced by the interaction of waves and currents in the ridge and runnel system at the south end of Jekyll Island, Georgia. (Knife for scale.)

C. Interference ripples preserved in sandstone (Paleozoic). Appalachian Valley and Ridge Province, northwest Georgia. (Rock hammer for scale.)

Figure 5.3 Interference ripples.

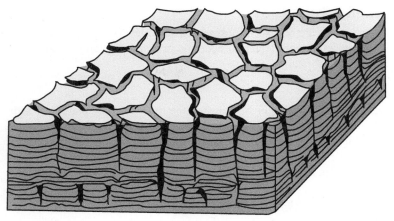

A. Mudcracks.

B. Mudcracks in clay-sized Recent (modern) sediment, Frederick, Maryland. (Pencil for scale).

Figure 5.4 Mudcracks.

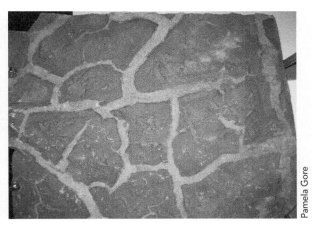

C. Mudcracks (Triassic), on the bottom of a sandstone bed, Hampshire. Formation, near Altoona, Pennsylvania, on display at Northern Virginia Community College.

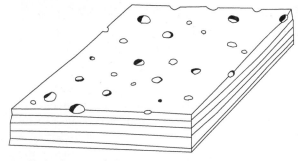

A. Raindrop prints.

Figure 5.5 Raindrop prints.

B. Raindrop prints and mudcracks, Bartow County, Georgia.

Internal Bedding Structures

Internal bedding structures are best seen looking at a side view of a sedimentary rock or sequence of sedimentary rocks. Internal bedding structures include stratification, graded beds, and cross stratification.

Stratification

Stratification (also called layering or bedding) is the most obvious feature of sedimentary rocks (Figure 5.6). The layers (or strata) are visible because of differences in the color or texture of adjacent beds. Strata thicker than 1 cm are commonly referred to as **beds**. Thinner layers are called **laminations** (Figure 5.7). The upper and lower surfaces of these layers are called **bedding planes**.

Figure 5.6 Stratification or bedding in sedimentary rocks (Paleozoic), Red Mountain road cut, Birmingham, Alabama.

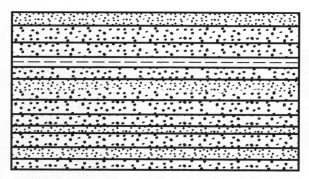

A. Laminations (layers thinner than 1 cm).

B. Laminations in beach sand, St. Simons Island, Georgia. (Pencil for scale.)

Figure 5.7

Varves are a special type of lamination, first described in glacial lake sediments, that form as a result of seasonal changes over the course of a year. Varves are generally graded, with the coarser material at the bottom (silt or sand) representing spring and summer melt water runoff, and the finer material at the top representing slow settling of clays and organic matter from suspension during winter months, when the lake is covered with ice. Counting of varves in the geologic record has been used to measure the ages of some sedimentary deposits.

Graded Bedding

Graded bedding results when a sediment-laden current (such as a **turbidity current**) begins to slow down (Figure 5.8). The grain size within a graded bed ranges from coarser at the bottom to finer at the top. The transition from one graded bed to the next is a crisp contact separating fine sediment below from coarse sediment above. Hence, graded beds look different right-side up than they do upside down, and therefore may be used as geopetal or paleo-up indicators.

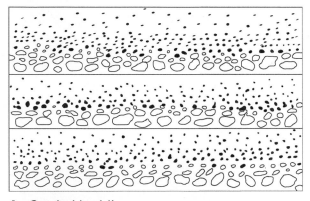

A. Graded bedding.

B. Graded bedding, New Jersey. (Pencil for scale.)

Figure 5.8 Graded bedding.

Cross Stratification

Cross stratification is a general term for the internal bedding structure produced in sand by moving wind or water (Figure 5.9). If the individual inclined layers are thicker than 1 cm, the cross stratification may be referred to as **cross bedding**. Thinner inclined layering is called **cross lamination**.

Cross stratification forms beneath ripples and dunes. The layering is inclined at an angle to the horizontal, dipping downward in the down-current direction. Hence, cross beds may be used as paleocurrent indicators, or indicators of ancient current flow directions. Cross beds usually curve at the bottom edge, becoming tangential to the lower bed surface. In contrast, the upper edge of individual inclined cross beds is usually at a steep angle to the overlying bedding plane. Hence, cross beds may also be used as geopetal or paleo-up indicators.

Sole Marks

Sole marks are bedding plane structures preserved on the bottom surfaces (or "sole") of beds. They generally result from the filling in of impressions made into the surface of soft mud by the scouring action of the current, or by the impacts of objects carried by the current. If sand is deposited later over the mud, filling in these structures, they will be preserved in relief on the bottom of the sandstone bed. (These structures are not usually seen on the surfaces of shale beds because they tend to weather away.) Sole marks include tool marks and flute marks.

Tool Marks

Tool marks are produced as "tools" (objects such as sticks, shells, bones, or pebbles), carried by a current, bounce, skip, roll, or drag along the sediment surface (Figure 5.10). They are commonly preserved on the lower surfaces of sandstone beds as thin ridges. Tool marks are generally aligned parallel to the direction of current movement.

A. Cross stratification in beach sand, Jekyll Island, Georgia. (Knife for scale.)

B. Cross stratification in sandstone (Late Paleozoic), Birmingham, Alabama. (Pencil for scale.)

C. Cross-bedded sandstone (Pennsylvanian), Cloudland Canyon, Georgia.

D. Large-scale cross bedding in sandstone (Triassic), Bay of Fundy, Nova Scotia, Canada. (Geology student for scale.)

Figure 5.9   Cross stratification.

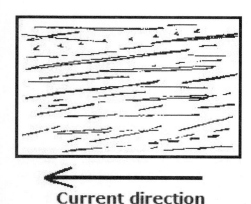

← **Current direction**

A. Tool marks.

Figure 5.10   Tool marks.

B. Tool marks in shale, Kentucky.

Flute Marks

Flute marks are produced by erosion or scouring of muddy sediment, forming scoop-shaped depressions (Figure 5.11). They are commonly preserved as bulbous or mammillary natural casts on the bottoms of sandstone beds. Because of their geometry, flute marks (also called flute casts) can be used to determine paleocurrent directions.

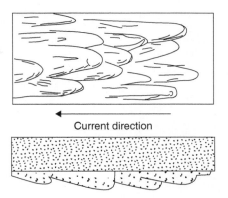

Current direction

Figure 5.11   Flute marks. *Top,* View of the bottom or "sole" of a sandstone bed, showing bulbous flute marks. Current flowed from right to left. *Bottom,* Side view of the sandstone bed with flute marks sticking out at the bottom.

ORGANIC OR BIOGENIC SEDIMENTARY STRUCTURES

Organic or biogenic sedimentary structures are those formed by living organisms interacting with the sediment. The organisms may be animals that walk on or burrow into the sediment, or they may be plants with roots that penetrate the sediment, or they may be bacterial colonies that trap and bind the sediment to produce layered structures (stromatolites).

Trace Fossils or Ichnofossils

Trace fossils or ichnofossils include tracks, trails, burrows, borings, and other marks made in the sediment by organisms. They are **bioturbation structures** formed as the activities of organisms disrupt the sediment. As organisms tunnel through sediment, they destroy primary sedimentary structures (such as laminations) and produce burrow marks. Bioturbation continuing over a long time thoroughly mixes and homogenizes the sediment. Through this process, laminated sediment can be altered to massive, homogeneous sediment without readily discernible layering.

Tracks

Tracks or footprints are impressions produced by the feet of animals (Figure 5.12). In some cases, tracks are found as sole marks on the bottoms of beds, where sediment has filled the tracks and preserved them as casts.

Tellus Science Museum

Figure 5.12   Dinosaur track. Cast of a *Tyrannosaurus rex* footprint, New Mexico, on display at Tellus Museum, Cartersville, Georgia. (Track is 33 inches long and 28 inches wide.).

Trackways

A **trackway** is a line of tracks showing the path along which an animal walked (as opposed to an isolated footprint) (Figure 5.13).

A. Raccoon trackway (recent), near Sanford, North Carolina.

B. Dog trackway (recent), near Wilson, North Carolina.

Figure 5.13 Trackways.

C. Dinosaur trackways, Dakota Group (Cretaceous), Dinosaur Ridge, near Morrison, Colorado.

A. *Climactichnites* trails in rippled sandstone (Late Cambrian), Quebec. Trails are about 10 to 20 cm wide. Specimen PAL532847 on display at the Smithsonian Institution, National Museum of Natural History.

Figure 5.14 Trails.

B. Trails in red siltstone (Triassic), Culpeper Basin, near Manassas, Virginia. (Scale in centimeters.)

Trails

Trails are groovelike impressions on the surface of a bed of sediment produced by an organism that crawls or drags part of its body (Figure 5.14). Trails may be straight or curved.

Burrows

Burrows are excavations made by animals into soft sediment (Figure 5.15). Burrows may be used by organisms for dwellings, or they may be produced as a subterranean organism moves through the soil or sediment in search of food. Burrows are commonly filled in by sediment of a different color or texture than the surrounding sediment, and in some cases, the burrows have an internally laminated backfilling. Burrow fillings can become cemented and hard, weathering out of the rock in ropelike patterns.

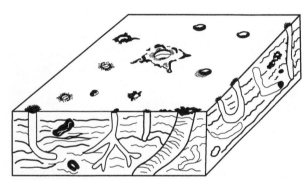

A. Several types of burrows, including branching, U-shaped, and vertical.

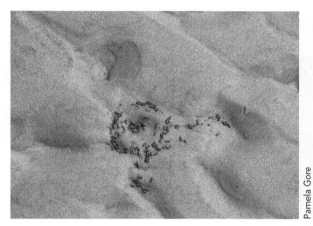

B. Burrow surrounded by fecal pellets in rippled sand, Georgia coast. Pellets are cylindrical and about 1 mm in diameter.

C. Branching burrows in siltstone (Triassic), Deep River Basin, North Carolina. (Lens cap for scale is about 5 cm in diameter.)

D. *Skolithos* worm burrows in quartzite stream cobbles (Cambrian), Henson Creek, Prince George's County, Maryland. (Scale in centimeters.)

Figure 5.15 Burrows.

Borings

Borings are holes made by animals into *hard material*, such as wood, shells, rock, or hard sediment (Figure 5.16). Borings are usually circular in cross section. Some snails are predators and produce borings or drill holes into other molluscs, such as clams, to eat them. Another mollusc, known as the shipworm, drills holes into wood. Sponges also produce borings, commonly riddling shells with numerous small holes.

Root Marks

Root marks are the traces left by the roots of plants in ancient soil zones (called paleosols) (Figure 5.17). Root marks typically branch downward in a pattern resembling an upside-down tree. Root marks are sometimes gray or greenish, penetrating reddish-brown paleosols. This contrast in color can make them easy to see and identify.

Pamela Gore

A. Boring in *Arca* bivalve shell, produced by carnivorous moon snail. (Scale in millimeters.)

Pamela Gore

B. Borings in bivalve shells, St. Augustine, Florida. (Largest shell is about 3 cm wide.)

Figure 5.16 Borings.

Pamela Gore

Figure 5.17 Root marks in siltstone (Triassic), Deep River Basin, near Sanford, North Carolina. (Rock hammer for scale.)

BIOSTRATIFICATION STRUCTURES

Biostratification structures are sedimentary layers produced through the activities of organisms. Stromatolites are the only type of biostratification structure we will study.

Stromatolites

Stromatolites are moundlike structures formed by colonies of sediment-trapping *cyanobacteria* (Figure 5.18). These organisms inhabit some carbonate tidal flats and produce domelike laminations in lime mud (fine-grained limestone or micrite). Stromatolites are organosedimentary structures and not fossils because they contain no recognizable anatomical features. They can, however, be considered to be trace fossils.

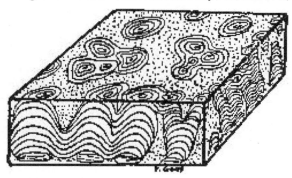

A. Stromatolites.

B. Stromatolite, 1.7 billion years old, from Wiluna, Western Australia, on display at Tellus Museum, Cartersville, Georgia.

C. Stromatolites in limestone, Maryland. (Scale in centimeters.)

D. Digitate (finger-like) stromatolites, Ordovician, western Maryland. (Finger for scale.)

E. Ordovician-age stromatolite, western Maryland.

Figure 5.18 Stromatolites.

Stromatolites form today in only a few places in the world, primarily in hypersaline environments (environments that have salinity in excess of sea water, preventing animal grazing), such as Shark Bay, Australia, or in freshwater carbonate-precipitating lakes. In the geologic record, most stromatolites are found in Precambrian and Lower Paleozoic limestones. After that time, they were preyed on by grazing organisms like snails. Cyanobacteria that formed these stromatolites were photosynthetic, and they are likely responsible for changing the character of Earth's atmosphere from one dominated by carbon dioxide to one with significant quantities of free oxygen.

DETERMINING UP DIRECTION

When you examine a sequence of beds that has been tectonically deformed and possibly overturned, it is necessary to determine the *up direction*. This is done by studying the sedimentary structures for clues (Figure 5.19).

Sedimentary structures such as graded beds, cross beds, mudcracks, flute marks, symmetrical (but not asymmetrical) ripples, stromatolites, burrows, tracks, and other structures can be used to establish the original orientation of the beds. (Fossils can also be used to establish up direction if they are present in the rock in life position.)

Carefully examine the sedimentary structures in *any* dipping sedimentary sequence, because the rocks can be *overturned* by tectonic forces, and what initially appears to be younger because it is on top may in fact turn out to be at the bottom of the section!

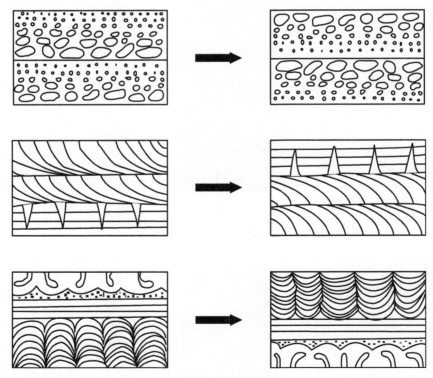

Figure 5.19 Illustration of overturned beds. *Left,* Right-side up. *Right,* Upside down.

Summary

Here is a list of the sedimentary structures mentioned in this lab.

I. Inorganic Sedimentary Structures
  A. BEDFORMS AND SURFACE MARKINGS
  1. Ripples
     - Asymmetrical ripples (including lingoid ripples)
     - Symmetrical ripples
     - Interference ripples
  2. Mudcracks
  3. Raindrop prints

  B. INTERNAL BEDDING STRUCTURES
  1. Stratification (strata)
     - Beds
     - Laminations or laminae
     - Varves
  2. Graded bedding
  3. Cross stratification
     - Cross bedding (cross beds)
     - Cross lamination

  C. SOLE MARKS
  1. Tool marks
  2. Flute marks

II. Organic or Biogenic Sedimentary Structures
  A. TRACE FOSSILS OR ICHNOFOSSILS
  1. Tracks
  2. Trackways
  3. Trails
  4. Burrows
  5. Borings
  6. Rootmarks

  BIOSTRATIFICATION STRUCTURES
  1. Stromatolites

Shells on the beach near St. Augustine, Florida.

Pamela Gore

Sedimentary Structures Exercises

PRE-LAB EXERCISES

1. Match the sedimentary structure with the environment in which it is most likely to be found. Put the letter in the blank.

1. _____	Mudcracks	a.	turbidity currents
2. _____	Stromatolites	b.	dried up lake
3. _____	Symmetrical ripples	c.	glacial lake
4. _____	Asymmetrical ripples	d.	tidal flat
5. _____	Graded bedding	e.	river
6. _____	Varves	f.	wave-washed shoreline

2. Which of the **sedimentary structures** in this lab may be useful in determining paleocurrent directions? (List four different structures.)

    _____   _____

    _____   _____

3. Which of the **sedimentary structures** in this lab may be useful in helping determine the top from the bottom of a bed (up indicators)? (List four different structures.)

    _____

    _____

    _____

    _____

4. Place an x in the table for the environments in which the sedimentary structure can form.

Sedimentary Structure	River	Shallow sea	Beach	Tidal flat	Dry lake bottom	Sand dunes (wind)	Deep Sea
Laminations							
Asymmetrical ripples							
Symmetrical ripples							
Mudcracks							
Raindrop prints							
Cross stratification							
Graded bedding							
Tracks							
Burrows							
Stromatolites							

5. Identify the following sedimentary structures.

a. _____

b. _____

Pamela Gore

c. _____

Pamela Gore

d. _____

LAB EXERCISES

Using the sedimentary structures provided in the lab, fill in the table below.

For each sample, identify the sedimentary rock. Use the illustrations in this lab to identify the sedimentary structure. Note that some lab specimens may have more than one sedimentary structure, such as both laminations and burrows, or burrows and ripple marks. Use the Summary page to determine whether the sedimentary structure formed through inorganic processes or biogenic processes. If there is more than one sedimentary structure, there may be more than one answer. Finally, using what you learned in this lab, try to interpret the environment in which the sedimentary structure(s) formed. If the structure forms in moving wind or water, indicate that and interpret whether wave or current motion may have been involved. The environment of deposition may have been a river or beach. If the structure forms as mud dries, the environment of deposition may have been a lake or tidal flat.

Sample	Rock Type	Name of Sedimentary Structure	Inorganic or Biogenic?	Interpret the environment of deposition
1				
2				
3				

Sample	Rock Type	Name of Sedimentary Structure	Inorganic or Biogenic?	Interpret the environment of deposition
4				
5				
6				
7				
8				
9				
10				
11				
12				
13				
14				
15				
16				
17				
18				

Depositional Sedimentary Environments

WHAT IS A SEDIMENTARY ENVIRONMENT?

A sedimentary environment is an area of Earth's surface where sediment is deposited. It can be distinguished from other areas on the basis of its *physical*, *chemical*, and *biological* characteristics. The unique processes operating in that environment leave a mark on the sediments deposited there. The characteristics of sedimentary rocks, including grain size, grain shape, sorting, composition, color, sedimentary structures, and fossils all provide clues to the environment in which they were deposited.

Before studying ancient sedimentary environments, it is helpful to consider the types of sedimentary environments present on Earth today.

TYPES OF SEDIMENTARY ENVIRONMENTS

Sedimentary environments can be grouped into three categories: continental environments (inland), transitional environments (along the coast), and marine environments (offshore) (Figure 6.1).

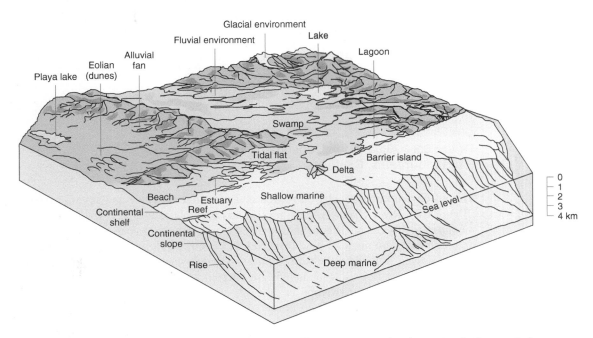

Figure 6.1   (From Levin, H., 2013, *The Earth Through Time* (10th edition), figure 5-2, p. 84. This material is reproduced with permission of John Wiley & Sons, Inc.)

Continental Environments

Continental environments are present on the continents (Figure 6.2), and include alluvial fans (fan-shaped deposit of sediment at the base of a mountain), fluvial environments (rivers), lacustrine environments (lakes), eolian (sometimes spelled *aeolian*) environments (deserts), and paludal environments (swamps).

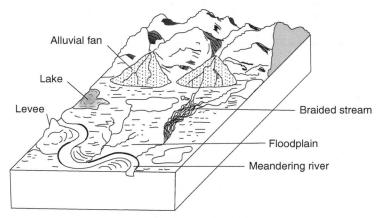

Figure 6.2 Continental sedimentary environments.

Alluvial fans are fan-shaped deposits of sediment formed at the base of mountains (Figure 6.3). Alluvial fans are most common in arid and semiarid regions, where rainfall is infrequent but torrential, slopes are steep, and erosion is rapid. Alluvial fan sediment is typically coarse, poorly sorted gravel and sand.

Figure 6.3 Alluvial fans, Joshua Tree National Park, Pinto Mountains, California.

Fluvial environments include **braided** and **meandering river** and stream systems (Figure 6.4). River channels, bars, levees, and floodplains are parts (or subenvironments) of the fluvial environment. Channel deposits consist of coarse, rounded gravel and sand. Bars are made of sand or gravel. A levee is an elongated mound along the side of a river, formed by the deposition of sediment when a river floods. Levees are made of fine sand or silt. Floodplains are covered by silt and clay, and they may be associated with lakes or swamps.

A. Meandering river, southeastern United States.

B. Meandering stream, South Fork of Peachtree Creek, Decatur, Georgia.

C. Sand bar along the Oconee River, Sandersville, Georgia.

D. Gravel bar along Henson Creek, Prince George's County, Maryland.

Figure 6.4 Fluvial environments.

Lakes (lacustrine environments) are diverse (Figure 6.5). They may be large or small, shallow or deep, fresh water or salt water, and filled with terrigenous, carbonate, or evaporite sediments. Sedimentary structures such as mudcracks, wave ripples, laminations, and varves may be present in lake sediments. Fine sediment and organic matter settling in some lakes produced laminated oil shales.

Deserts (eolian environments) are areas with little or no rainfall during the year (Figure 6.6). Deserts usually contain vast areas where sand is deposited in dunes. Dune sands are well-sorted and well-rounded, and sand grains are frosted or polished, without associated gravel or clay. Cross bedding is common.

Swamps (paludal environments) are areas of standing water with trees (Figure 6.7). Decaying plant matter accumulates to form peat, which can eventually become coal. Swamps may be present on floodplains associated with fluvial environments.

A. Bonneville Salt Flats, Utah. This area is a dried lake bed, near the Great Salt Lake, covered in white rock salt. The lake was once more than 1,000 feet deep and nearly the size of Lake Michigan during the last Ice Age, about 20,000 years ago. As the climate changed, the lake dried up.

B. Aerial view of a dry playa lake, with white salt deposits, California. Playa lakes are ephemeral lakes that dry out seasonally.

C. Lake Eliza, a shallow hypersaline lake in South Australia. Hypersaline lakes have salinity greater than that of sea water.

D. Ephemeral pond near the Great Salt Lake, Utah.

Figure 6.5 Lacustrine (lake) environments.

E. Jones Lake State Park, North Carolina. Jones Lake is a type of freshwater lake called a Carolina Bay (named for the bay tree). Many such lakes are present on the Atlantic Coastal Plain.

A. Relict Pleistocene dune sand along the Flint River in Albany, Georgia.

B. Mesquite flat sand dunes in Death Valley National Park.

Figure 6.6 Ancient and modern eolian (desert) deposits.

A. Cypress swamp near Albany, Georgia.

B. Okefenokee Swamp, Waycross, Georgia.

Figure 6.7 Paludal (swamp) environments.

Transitional Environments

Transitional environments are at or near the coast, where the land meets the sea (Figure 6.8). Transitional sedimentary environments include deltas, beaches and barrier islands, lagoons, salt marshes, and tidal flats. Tidal flats are low-lying areas that are alternately covered by water and exposed to the air each day as the tides rise and fall.

Deltas are fan-shaped deposits of sediment, formed where a river flows into a standing body of water, such as a lake or sea (Figure 6.9). Coarser sediment (sand) tends to be deposited near the mouth of the river; finer sediment is carried seaward and deposited in deeper water. Well-known deltas include the Mississippi River delta and the Nile River delta.

Beaches and barrier islands are shoreline deposits exposed to wave energy and dominated by sand with a marine fauna (Figure 6.10). Barrier islands are separated from the mainland by a lagoon. They are commonly associated with **tidal flat** deposits.

Lagoons are bodies of water on the landward side of barrier islands. They are protected from the pounding of the ocean waves by the barrier islands, and they contain finer sediment than the beaches (usually silt and mud). Lagoons are also present behind reefs or in the center of atolls.

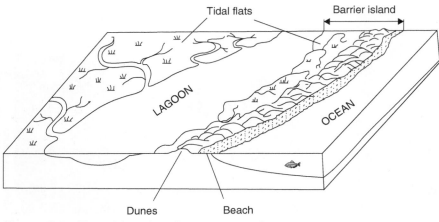

Figure 6.8   Transitional sedimentary environments.

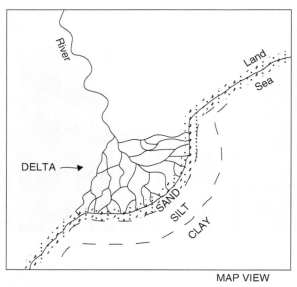

A. Diagram of a delta.

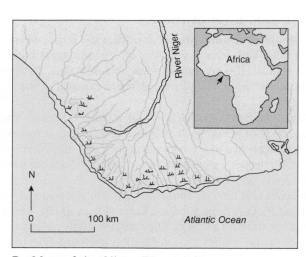

B. Mississippi River delta (satellite imagery).

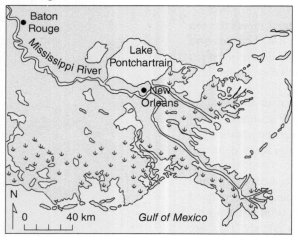

C. Map of the Mississippi River delta.
(From Levin, H., 2013, *The Earth Through Time*
(10ᵗʰ edition), figure 8-9, p. 88. This material is
reproduced with permission of
John Wiley & Sons, Inc.)

D. Map of the Niger River delta.
(From Levin, H., 2013, *The Earth Through Time*
(10ᵗʰ edition), figure 8.9, p. 88. This material is
reproduced with permission of
John Wiley & Sons, Inc.)

Figure 6.9   Deltas.

A. Outer Banks of North Carolina, a barrier island. Kitty Hawk, North Carolina. Note the lagoon (Pamlico Sound) in the background at right, with the mainland barely visible on the horizon. The tan area of sand in the background at right is Jockey's Ridge, the tallest natural sand dune system in the eastern United States.

B. A narrow part of the barrier island near Cape Hatteras, North Carolina. The lagoon (Pamlico Sound) is on the left, and the Atlantic Ocean is on the right.

C. The sand is white at Pensacola Beach, on the Florida Gulf coast.

D. Coquina outcrops on the beach at Washington Oaks State Park, Florida, Atlantic coast.

Figure 6.10 Beaches and barrier islands.

Tidal flats are areas that are periodically flooded and drained by the tides (usually twice each day) (Figure 6.11). Tidal flats are areas of low relief, commonly cut by meandering tidal channels. Laminated or rippled clay, silt, and fine sand (either terrigenous or carbonate) may be deposited. Burrows and tracks are common. Stromatolites may be present on carbonate tidal flats, if conditions are appropriate (high salinity).

Salt marshes, associated with some tidal flats, also experience tidal fluctuation (Figure 6.12). Salt marshes are covered by salt-tolerant grasses, and the muddy sediment is heavily bioturbated (burrowed and root-marked). Oyster shells are common. Salt marshes are present in some areas along the Atlantic, Gulf and Pacific coasts.

A. Low tide at Five Islands Provincial Park, Bay of Fundy, Nova Scotia, Canada.

B. High tide at Five Islands Provincial Park, Bay of Fundy, Nova Scotia, Canada.

Figure 6.11 Tidal flats.

C. Tidal flat with ripples at low tide, Wassaw Island, Georgia coast.

Marine Environments

Marine environments are offshore in the seas or oceans. Marine environments include reefs, the continental shelf, slope, rise, and abyssal plain.

Reefs are wave-resistant, mound-like structures made of the calcareous skeletons of organisms such as corals and certain types of algae (Figure 6.13). Most modern reefs are in warm, clear, shallow seas at tropical latitudes (between 30°N and 30°S of the equator). **Atolls** are ringlike reefs surrounding a central lagoon.

Sunlight is required for reef growth because of the presence of symbiotic algae called **zooxanthellae** that live in the tissues of corals.

A. Salt marsh at low tide, with *Spartina* grass and oyster shells on mud, Jekyll Island, Georgia.

B. Salt marsh at high tide, Tybee Island, Georgia.

Figure 6.12   Salt marshes.

A. Aerial view of Florida Reef from the Gulf Stream side of the reef.

B. Oblique color shaded-relief view and bathymetric profile of Alabama Alps Reef, Pinnacles Region, Gulf of Mexico.

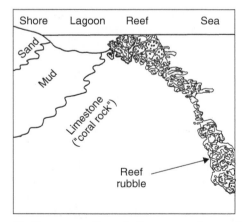

C. Cross-sectional diagram of a reef.

Figure 6.13   Reefs

D. Live scleractinian coral with other reef-dwelling organisms.

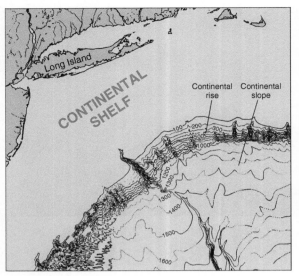

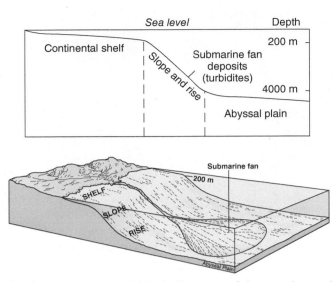

A. Map of the continental shelf, slope and rise, south of Long Island, New York. (Stowe, K., *Ocean Science*, 1979, Figure 3.6, p.84. This material is reproduced with permission of John Wiley & Son, Inc.)

B. Cross section and block diagram of the continental shelf, slope and rise, and abyssal plain. Submarine fans can form on the continental rise at the mouth of submarine canyons. (From Levin, H., 2013, *The Earth Through Time* (10th edition), figure 5-5, p. 86. This material is reproduced with permission of John Wiley & Sons, Inc.)

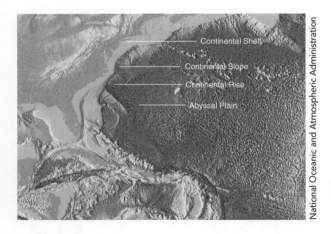

C. Relief on the floor of the Atlantic Ocean off the east coast of North America, showing continental shelf, slope and rise, and abyssal plain.

Figure 6.14 Continental shelf, slope, rise and abyssal plain.

The **continental shelf** is the flooded edge of the continent (Figure 6.14). It is relatively flat (with a slope of less than 0.1°) and shallow (less than 200 m or 600 ft deep), and it may be hundreds of miles wide. Continental shelves are exposed to waves, tides, and currents and are covered by sand, silt, mud, and gravel. The flooding of the edges of the continents occurred when the glaciers melted at the end of the last Ice Age, about 10,000 years ago. The flatness of the continental shelf is due in part to erosion by waves at sea level low stands during glaciations and in part to sediment deposition.

The continental slope and continental rise are located seaward of the continental shelf. The **continental slope** is the 5°–25° "drop-off" at the edge of the continental shelf. The continental slope passes seaward into the **continental rise**, which has a more gradual slope. Thick accumulations of sediment are present on the continental rise.

During the Pleistocene Ice Ages, when sea level was much lower than today, the continental shelves were exposed as dry land and cut by river channels. In some places, the catastrophic flow of meltwater from glacial lakes carved deep submarine canyons into the edge of the continental slope. **Turbidity currents** flowing down these canyons deposited their sediment to form huge **submarine fans** at the base of the slope, as part of the continental rise. Turbidity current deposits or **turbidites** are characterized by graded bedding. Locally, submarine landslides down the continental slope have produced thick sediment deposits at the base of the slope, which are part of the continental rise. Submarine landslides can trigger tsunami.

The **abyssal plain** is the deep ocean floor. It is basically flat and is covered by very fine grained sediment, deposited over an irregular basaltic ocean crust with volcanic peaks. The sediment consists primarily of clay and the shells of microscopic organisms such as **foraminifera**, **radiolarians**, **diatoms**, and **coccolithophores** (see Laboratory 9: Microfossils and Introduction to the Tree of Life). Abyssal plain sediments can include chalk, diatomite, and shale. The clays are typically red to brown and can contain fish bones and teeth.

FEATURES THAT HELP US IDENTIFY ANCIENT SEDIMENTARY ENVIRONMENTS

Sedimentary rocks contain clues that help us to determine the sedimentary environment in which they were deposited millions of years ago. By an examination of the *physical*, *chemical*, and *biological* characteristics of the rock, we can interpret the environment of deposition. Each sedimentary environment has its unique combination of *physical*, *chemical*, and *biological* features. These features help us to identify the environment where a sediment was deposited.

In lab, you will be examining hand specimens of sedimentary rocks, describing their *physical*, *chemical*, and *biological* features, and interpreting their possible sedimentary environments of deposition. Geologists consider the characteristics that we will study in lab (Box 6.1), but they also study the *geometry* of the sedimentary deposits, the *vertical sequence* of sedimentary rocks, and the *paleocurrent directions* as determined from sedimentary structures such as asymmetrical ripples, tool marks, or flute marks (see Laboratory 5: Sedimentary Structures). In places, one environment grades into another; for example, a fluvial environment grades into a deltaic environment.

Certain generalizations can be made that help in identifying the depositional environment. For example, *fluvial sequences become finer upward*, whereas *delta and lacustrine sequences coarsen upward*. These predictable changes occur because the environments migrate over one another as sea level (or lake level) changes or as a basin fills with sediment.

As a general rule, *grain size is coarser in shallow-water high-energy environments* where waves or currents are active. Waves and currents transport finer sediment offshore into low-energy environments, generally in deep, quiet water. *Fine grain size indicates deposition in a low-energy, quiet-water environment.*

In some areas far from shore (or far from a source of terrigenous input) only the shells of planktonic microorganisms contribute to the sediment. These microscopic shells accumulate to form rocks such as chalk or diatomite.

Box 6.1   Characteristics of rocks that can be used to interpret the depositional environment

Physical Characteristics

A. Rock type
B. Texture
   1. Grain size
   2. Grain shape
   3. Sorting
C. Composition of detrital grains
   1. Terrigenous grains (quartz, feldspar, rock fragments)
   2. Carbonate grains (shells, oolites, etc.)
   3. Organic material
D. Sedimentary structures

Chemical Characteristics

A. Grain composition (minerals present)
B. Type of cement
C. Rock color
D. Presence of evaporites

Biological Characteristics

A. Fossils
   1. Shells
   2. Bones and teeth
   3. Plant material
B. Traces of organisms
   1. Fecal pellets or coprolites
   2. Biogenic sedimentary structures
   3. Organic matter

Tables 6.1, 6.2, and 6.3 give a general guide to interpreting the depositional environment of sedimentary rocks. This is not meant to give exhaustive coverage of all possible environments. For example, freshwater and salt lakes are grouped together under "lacustrine," but a freshwater lake generally would not contain evaporites. Variations in types of lake (or other environments) are due to factors such as climate or tectonic setting. As a result, a particular sedimentary environment might only have a few of the characteristics listed in the table.

In addition, some environments have a number of distinctive subenvironments that are not distinguished separately here. For example, fluvial environments have floodplains, channels, point bars, levees and other subenvironments. Interpretation of subenvironments is beyond the scope of this course.

Some common rock types, such as shale or sandstone, are present in numerous environments. By examining other features, such as sedimentary structure or fossils, you may be able to increase the certainty of the interpretation.

TABLE 6.1   CONTINENTAL SEDIMENTARY ENVIRONMENTS

Characteristic	Alluvial Fan	Fluvial (River)	Lacustrine (Lake)	Desert/Eolian (Dunes)	Paludal (Swamp)
Rock type	Breccia, conglomerate, arkose	Conglomerate, sandstone, siltstone, shale	Siltstone, shale, limestone, or evaporites (gypsum, rock salt)	Quartz arenite (sandstone) or gypsum	Peat, coal, black shale, siltstone
Composition	Terrigenous	Terrigenous	Terrigenous, carbonate, or evaporite	Terrigenous or evaporite	Terrigenous
Color	Brown or red	Brown or red	Black, brown, gray, green, red	Yellow, red, tan, white	Black, gray, or brown
Grain size	Clay to gravel	Clay, silt, sand, gravel (fining upward)	Clay to silt or sand (coarsening upward)	Sand	Clay to silt
Grain shape	Angular	Rounded to angular	—	Rounded, polished, or frosted grains	—
Sorting	Poor	Variable	Variable	Good	Variable
Inorganic sedimentary structures	Cross bedding and graded bedding	Asymmetrical ripples, cross bedding, graded bedding, tool marks	Symmetrical ripples, lamination, cross bedding, graded bedding, mudcracks, raindrop prints	Cross bedding	Laminated to massive
Organic or biogenic sedimentary structures	—	Tracks, trails, burrows	Tracks, trails, burrows, rare stromatolites	Tracks, trails	Root marks, burrows
Fossils	—	Rare freshwater shells, bones, plant fragments	Freshwater shells, fish, bones, plant fragments	—	Plant fossils, freshwater shells, bones, fish

TABLE 6.2 TRANSITIONAL SEDIMENTARY ENVIRONMENTS

Characteristic	Delta	Barrier Beach	Lagoon	Tidal Flat
Rock type	Sandstone, siltstone, shale, coal	Quartz arenite, coquina	Siltstone, shale, limestone, oolitic limestone or gypsum	Siltstone, shale, calcilutite, dolostone or gypsum
Composition	Terrigenous, some organic	Terrigenous or carbonate	Terrigenous, carbonate, or evaporite	Terrigenous, carbonate, or evaporite
Color	Brown, black, gray, green, red	White to tan	Dark gray to black, brown	Gray, brown, tan
Grain size	Clay to sand (coarsening upward)	Sand	Clay to silt	Clay to silt
Grain shape	—	Rounded to angular	—	—
Sorting	Variable	Good	Poor	Variable
Inorganic sedimentary structures	Cross bedding, graded bedding	Cross bedding, symmetrical ripples	Lamination, ripples, cross bedding	Lamination, mudcracks, ripples, cross bedding
Organic or biogenic sedimentary structures	Trails, burrows	Tracks, trails, burrows	Trails, burrows	Stromatolites, trails, tracks, burrows
Fossils	Plant fragments, shells	Marine shells	Marine shells	Marine shells

TABLE 6.3   MARINE SEDIMENTARY ENVIRONMENTS

Characteristic	Reef	Continental Shelf	Continental Slope and Rise	Abyssal Plain
Rock type	Fossiliferous limestone	Sandstone, shale, siltstone, fossiliferous limestone, oolitic limestone	Litharenite, siltstone, and shale (or limestone)	Shale, chert, micrite, chalk, diatomite
Composition	Carbonate	Terrigenous or carbonate	Terrigenous or carbonate	Terrigenous or carbonate
Color	Gray to white	Gray, brown, tan, green	Gray, green, brown	Black, white red
Grain size	Variable, frameworks, few to no grains	Clay, silt, sand, gravel	Clay to sand	Clay
Grain shape	—	—	—	—
Sorting	—	Poor to good	Poor	Good
Inorganic sedimentary structures	—	Lamination, cross bedding	Graded bedding, cross bedding, lamination, flute marks, tool marks (turbidites)	Lamination
Organic or biogenic sedimentary structures	—	Trails, burrows	Trails, burrows	Trails, burrows
Fossils	Corals, marine shells	Marine shells, shark teeth	Marine shells, rare plant fragments	Marine shells (mostly microscopic)

Quartz pebble conglomerate in a hoodoo or pedestal rock. Lookout Mountain, Georgia.

Depositional Sedimentary Environments Exercises

PRE-LAB EXERCISES

1. Match the rock feature (left column) with its possible depositional environment (right column).

    _____ Thin bedded shale   a. Fluvial

    _____ Tan cross-bedded sandstone, rounded grains, well-sorted   b. Lagoon or shallow sea with arid climate

    _____ Red beds with fining-upward sequences, asymmetrical ripples   c. Carbonate tidal flat

    _____ Coal and siltstone with plant fossils   d. Deep, quiet water offshore

    _____ Stromatolitic and intraclastic limestone   e. Windblown desert

    _____ Laminated evaporite minerals   f. Delta swamp

2. List *two* specific characteristics that you could use to distinguish between marine (ocean or salt water) and nonmarine (freshwater) sedimentary deposits. (Look at Box 6.1, and look at Tables 6.1, 6.2, and 6.3.)

    _____   _____

3. What are two of the best sedimentary structures that indicate an environment that is *periodically dry*? (Do not list sedimentary environments here.)

    _____   _____

4. List four high-energy sedimentary environments (i.e., those with currents, waves, or wind). (Do not list sedimentary structures or rocks here.)

    _____   _____

    _____   _____

5. List three low-energy sedimentary environments (i.e., those with calm water—no currents or waves). (Do not list sedimentary structures or rocks here.)

    _____   _____

    _____

6. List four sedimentary structures that form in high-energy environments. (Do not list sedimentary rocks or environments here.)

    _____   _____

    _____   _____

7. List four sedimentary structures that form in low-energy environments. (Do not list sedimentary rocks or environments here.)

_____   _____

_____   _____

LAB EXERCISES

Materials needed:

- Sedimentary rock samples illustrating features of major sedimentary environments, some of which have fossils, sedimentary structures, or multiple lithologies to form a suite of rocks
- Sedimentary environments classification charts in this lab
- Tools needed to identify sedimentary rocks (hand lens, grain size comparison chart, and dilute hydrochloric acid if available)

Instructions

1. The table below contains descriptions of actual sedimentary rocks and suites of associated rocks. Use the three sedimentary environment tables (Tables 6.1, 6.2, and 6.3) to identify *all possible sedimentary environments* for each sample. For some of these, you will have to list several possible environments. Certain sedimentary structures (or other features) may be present *in the environment* but might not be in a particular sample (perhaps because of small sample size). By examining other features, such as sedimentary structures or fossils, you may be able to increase the certainty of your interpretation. Use Laboratory 10: Invertebrate Macrofossils and Classification of Organisms or a fossil identification book if you are unfamiliar with any fossils listed here (such as brachiopods, bryozoans, or molluscs).

Sample	Rock Type	Sedimentary Structures	Chemical Characteristics	Biological Characteristics	List All Possible Sedimentary Environments
1	Quartz sandstone	None visible in this sample	Quartz grains and cement, white to tan color	Brachiopod fossil molds	
2	Fossiliferous limestone: very coarse, no mud	None visible in this sample	All calcium carbonate, white to gray color	Rock is dominated by very coarse coral pieces	
3	Fossiliferous limestone (fossils in a lime mud matrix)	Vague layering	All calcium carbonate, gray color	Rock is filled with whole and broken brachiopod shells and branching bryozoan colonies	

Sample	Rock Type	Sedimentary Structures	Chemical Characteristics	Biological Characteristics	List All Possible Sedimentary Environments
4	Gypsum with very thin laminations of calcium carbonate	Laminations	Evaporite minerals White color (due to gypsum), with gray calcium carbonate laminations	No fossils	
5	Breccia Poorly sorted, some clasts up to 6 inches (15 cm) in diameter Silty matrix	None visible in this sample	Red color due to iron oxide cement Grains are fragments of various types of metamorphic rocks	No fossils	
6	Limestone and dolostone, interlayered	Irregular layers and burrows	Carbonate minerals	Burrows	
7	Black shale Chert Chalk Very fine grained	Laminations in shale	Organic-rich black shale Gray to black chert White to gray chalk (calcium carbonate)	Microfossils probably present in the chalk although they were too small to identify in lab	
8	Limestone Fossiliferous shale Graywacke	Laminations, graded bedding	Gray limestone and shale, greenish gray graywacke	Fossils in shale and limestone, primarily molds of brachiopods	
9	Quartz sandstone (some are pebbly and poorly sorted) Micaceous siltstone	Pebbly sandstone is poorly layered Quartz sandstone is laminated and cross bedded No structures in siltstone	Reddish brown color due to iron oxides Quartz, muscovite, clay minerals	No fossils	

Sample	Rock Type	Sedimentary Structures	Chemical Characteristics	Biological Characteristics	List All Possible Sedimentary Environments
10	Coal Gray to black shale Greenish graywacke Tan to brown siltstone Red quartz sandstone	Quartz sandstone is cross bedded	Red color due to iron oxides in some samples Other samples gray, green, or black (organic matter from plants) Quartz	Siltstone and shale have plant fossils resembling ferns	
11	Quartz sandstone, well-sorted	Bedded and cross bedded	Red, orange, and tan quartz	No fossils	
12	Fossiliferous limestone, almost a coquina; some shells broken	None visible in this sample	Tan to light gray Carbonate	Full of fossils, primarily molluscs (bivalves and gastropods)	
13	Oolitic limestone	None visible in this sample	Gray to tan Carbonate	Some small marine shell fragments	

2. Examine the rock specimens provided by the instructor and fill in the table below. Use the three sedimentary environment tables (Tables 6.1, 6.2, and 6.3) to identify *all possible sedimentary environments* for each rock specimen. Some specimens might have more than one possible sedimentary environment. Some specimen trays might contain a suite of *several rock types* that tend to be associated together within an environment. Consider *all* of the samples in the suite when making your environmental interpretation.

Sample	Rock Type	Sedimentary Structures	Chemical Characteristics	Biological Characteristics	Possible Sedimentary Environments
1					
2					
3					
4					
5					

Sample	Rock Type	Sedimentary Structures	Chemical Characteristics	Biological Characteristics	Possible Sedimentary Environments
6					
7					
8					
9					
10					
11					
12					
13					
14					
15					

OPTIONAL ACTIVITY

Take a field trip near your campus (or in your local area) to a stream, lake, or beach to observe the types of sediment being deposited and look for sedimentary structures.

Pamela Gore

Pennsylvanian-age fluvial channel-fill sandstone, Cloudland Canyon State Park, Lookout Mountain, Georgia.

Stratigraphy and Lithologic Correlation

In this lab you will learn about sequences of sedimentary rocks and how they may be **correlated**, or traced between outcrops. Ideally, the rocks are correlated directly by walking along the contacts between adjacent rock units, across the countryside. This is seldom the case, however, particularly where vegetation and soil cover make rock exposures poor (as in humid areas, such as the eastern United States). In other situations, geologists are interested in beds deep below the Earth's surface, and they must use drill hole and core data to correlate the rocks.

Geologists study rocks in **outcrops** (natural or human-made exposures such as road cuts or quarries) or in drill cores. When studying an outcrop of sedimentary rock, the most obvious feature is **bedding** (also called **strata** or **layers**) (Figure 7.1). Although the rocks may be tilted or folded, the sediments were originally laid down in horizontal beds that extended as continuous layers in all directions (such as a layer of mud on the sea floor), with the oldest layers on the bottom and the youngest layers on top. A sequence of sedimentary rocks may be divided into a number of **lithostratigraphic units** of various sizes.

Pamela Gore

Figure 7.1 Providence Canyon, Georgia.

LITHOSTRATIGRAPHIC UNITS

A lithostratigraphic unit is defined as a body of sedimentary, extrusive igneous, metasedimentary, or metavolcanic strata that is distinguished on the basis of lithologic characteristics and stratigraphic position (i.e., position in the rock sequence).

The smallest lithostratigraphic rock unit is the **bed. Formations** are the *fundamental units of stratigraphy.*

By definition, formations are:

- Lithologically homogeneous (all beds are the same rock type or a distinctive set of interbedded rock types).

- Distinct and different from adjacent rock units above and below.

- Traceable from place to place across the countryside, and of sufficient thickness to be *mappable*. (Formations are commonly hundreds of feet thick, but they may be thinner or thicker).

Formations have formal *names* (Box 7.1). Formations are usually named for some geographic locality where they are particularly well exposed. (This locality is referred to as the *type section*.) If the beds are dominated by a single rock type, this might appear in the name. (Also, to be valid, the name of a formation must be published in the geological literature.) A set of similar or related formations is called a **group**. Groups also have names.

Box 7.1 Names of Formations and Groups

Examples of formation names are:

Chattanooga Shale
Wasatch Formation
Navajo Sandstone
Niobrara Chalk
Lockport Dolomite
Morrison Formation
Fort Payne Chert
Green River Formation
Fountain Arkose
Red Mountain Formation

Examples of group names are:

Knox Group
Mesaverde Group
Glen Canyon Group
Cincinnati Group
Chilhowee Group
Great Smoky Group
Pocono Group
Chesapeake Group
Dakota Group
Hamilton Group

Subdivisions within formations are called **members**. Members also have names. A formation, however, does not have to contain members. Members may be designated to single out units of special interest or economic value, such as coal beds or volcanic ash layers.

Boundaries of lithostratigraphic units are placed where the lithology (or rock type) changes. They may be placed at a distinct **contact**, or they may be set arbitrarily within a zone of gradation.

Virtually all lithostratigraphic units are time transgressive or **diachronous**, meaning that they, or their contacts, *cut across time lines*. For example, a particular unit of sandstone is early Cambrian in southern California and Nevada, but traced to the northeast to Colorado, it is late Cambrian, or roughly 35 million years younger. This is because the sandy beach environment migrated eastward as sea level changed over millions of years.

The lithostratigraphic terms (bed, member, formation, and group) refer to sedimentary, volcanic, metasedimentary, and metavolcanic rocks *only*.

Intrusive igneous rocks and **highly deformed and metamorphosed rocks** are called *lithodemic units*. For example, the rocks in the Piedmont are mostly metamorphic and intrusive igneous rocks and should therefore be called lithodemic units. The fundamental lithodemic unit is the **lithodeme** (roughly equivalent to a formation). The term *formation* should not be used for intrusive and metamorphic rock (according to the North American Stratigraphic Code of 1983, revised and published 2005; http://www.agiweb.org/nacsn/code2.html). The name of a lithodeme is a location plus a lithologic term, such as:

Pikes Peak Granite

Zoroaster Granite

Columbia River Basalt

Vishnu Schist

Baltimore Gneiss

Wissahickon Schist

Manhattan Schist

In this lab, we are primarily concerned with sequences of sedimentary rocks and sometimes with interbedded lava flows, so we will be using lithostratigraphic terminology.

STRATIGRAPHIC SECTIONS

Geologists study sequences of sedimentary rocks on a bed-by-bed basis. They measure the **thickness** of each bed, record the physical, chemical, and biological characteristics of the rock, and note the nature of the **contacts** (or bedding planes) between beds. Using these data, the geologist draws up a stratigraphic section for a particular sequence of rock. A **stratigraphic section** is a graphical or pictorial representation of the sequence of rock units. Standard symbols (called lithologic symbols) are used to refer to each rock type (Table 7.1).

TABLE 7.1 LITHOLOGIC SYMBOLS

Symbol	Lithology	Symbol	Lithology
	Breccia		Limestone
	Conglomerate		Dolostone
	Sandstone		Volcanic rocks
	Siltstone		Plutonic rocks
	Shale		Plutonic rocks
	Coal		Metamorphic rocks

Drawing a Stratigraphic Section

To draw a stratigraphic section, you must have data from a sequence of rocks (such as an outcrop, road cut, or core). You will need to have data on the thickness of each bed and on the major physical, chemical, and biological characteristics of that bed, as well as the character of its upper and lower contacts.

Before you start to draw your section, you need to examine your data to determine the total thickness of the section you plan to draw. Then, determine a proper scale so that the entire section will fit on your paper (such as, 1" = 100').

Draw a vertical column, where you will plot your data, and then mark off the thickness of each bed or formation using the scale you established. Draw in the contacts between units; if the contacts are erosional, draw a wavy line. After you have drawn in contacts, then draw in the lithologic symbols for each unit. Information on fossils, sedimentary structures, and so on may be placed within the unit or beside it using a special symbol or small sketch. Rock color may be illustrated with a special symbol or by coloring your section. Standard lithologic symbols have been established by oil companies and core logging companies, and you may use those or create your own.

Once you have drawn several stratigraphic sections for an area, you may begin to correlate them.

Lithologic Correlation

Geologists can draw **stratigraphic sections** for several outcrops (or cores) in an area and then trace beds from one section to another. This is called **lithologic correlation** (Figure 7.2). Basically, correlation demonstrates the equivalency of rock units across an area. The sections being correlated may be miles apart. Basically, a correlation is a hypothesis that units in two widely separated sequences are equivalent. Clearly, the more unique characteristics that two sections share, the greater the probability there is that the correlation is correct.

Correlation may be performed in several ways. Distinctive beds (called key beds or marker beds), distinctive sequences of beds, bed thicknesses, and unconformities may be traced between sections.

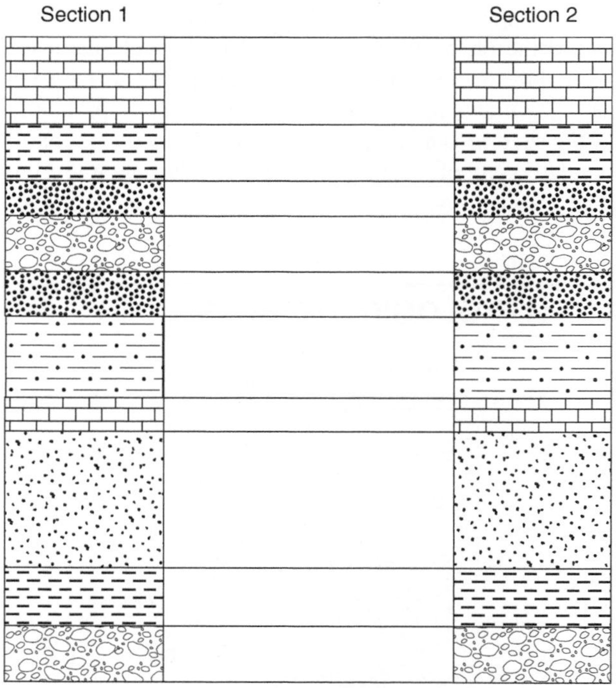

Figure 7.2 Illustration of lithologic correlation.

Key beds or **marker beds** tend to have some unusual, distinguishing feature that allows them to be readily identified (such as a bed of volcanic ash in a sedimentary sequence, or a bed of conglomerate in a sandstone sequence, or a bed of fossil shells or bones, or a bed of limestone in a shale sequence). It is helpful if key beds or marker beds are laterally extensive, which will aid in correlation over a large area.

Distinctive sequences of beds are also useful in correlation. For example, the sequence limestone—dolostone—limestone may be found within a thick unit of shales and siltstones and correlated between sections.

In some cases, beds can be correlated between sections based on their thicknesses. One of the best examples of this is the correlation of laminations in cores from the evaporites of the Castile Formation in the Permian of western Texas and New Mexico. Cores were drilled about 9 miles apart, and the thickness of the laminations matches almost exactly.

Unconformities

Sometimes, one or more rock units are missing from the middle of a sequence. Close examination of the outcrop shows a sharp or irregular contact where the missing rocks should be. This contact is called an **unconformity** (Figure 7.3). Unconformities are surfaces that represent a gap in the geologic record because of either erosion or nondeposition. Unconformities can be traced between stratigraphic sequences miles apart. Although unconformities can truncate rocks of many different ages, the sediments directly overlying the unconformity are roughly the same age.

A time of nondeposition or erosion might occur if sea level drops, exposing marine sediments to the air (subaerial exposure), followed by a later rise in sea level and renewed deposition. Or the sedimentation rate might decrease due to a time of drought in the source area, resulting in less sediment being supplied to the depositional basin.

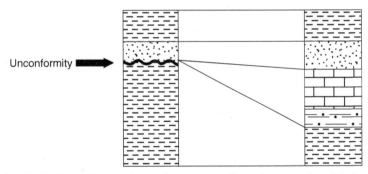

Figure 7.3   Illustration of an unconformity causing beds to be missing from a sequence.

There are four basic types of unconformities:

1. Angular unconformities
2. Nonconformities
3. Disconformities
4. Paraconformities

Angular Unconformities

Angular unconformities are characterized by an erosional surface that truncates folded or dipping (tilted) strata (Figure 7.4). Overlying strata are deposited basically parallel with the erosion surface.

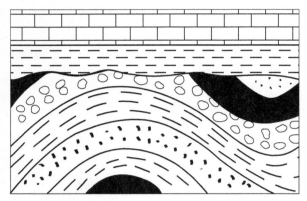
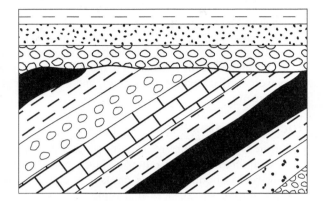

A. Angular unconformities.

Peter L. Kresan Photography

B. Angular unconformity at Siccar Point, Scotland.

Figure 7.4 Angular unconformities.

Nonconformities

Nonconformities are characterized by an erosional surface that truncates igneous or metamorphic rocks (Figures 7.5 and 7.6).

Disconformities

Disconformities are characterized by an irregular erosional surface that truncates flat-lying sedimentary rocks (Figure 7.7).

Paraconformities

Paraconformities are characterized by a surface of nondeposition separating two parallel units of sedimentary rock that is virtually indistinguishable from a sharp conformable contact; there is no obvious evidence of erosion (Figure 7.8). An examination of the fossils shows that there is a considerable time gap represented by the surface.

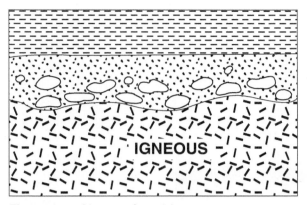 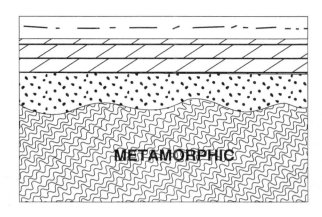

Figure 7.5   Nonconformities.

Pamela Gore

Figure 7.6   Photo of a nonconformity (black line), with rounded gravel overlying weathered metamorphic rock, Ellijay, Georgia.

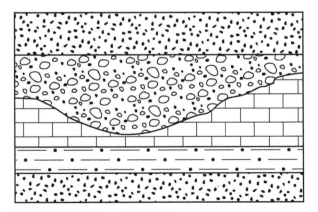 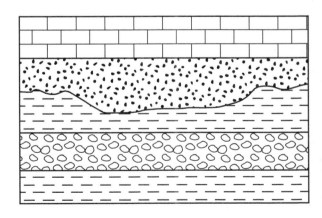

Figure 7.7   Disconformities.

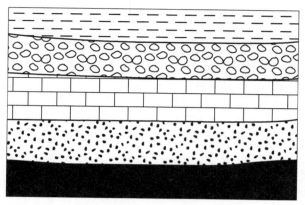

Figure 7.8   Paraconformity.

SEDIMENTARY FACIES

A **facies** is a unit of sedimentary rock deposited in a particular sedimentary environment. A facies has distinctive physical, chemical, and biological characteristics that serve as clues that help the geologist to interpret the environment where the rock was deposited. (Examples of sedimentary environments include beaches, rivers, lakes, deserts, alluvial fans, deltas, reefs, lagoons, and tidal flats.) You might refer to a red sandstone facies, or a mud cracked limestone facies, or a fossiliferous black shale facies.

Lateral Facies Changes

Beds can change laterally in thickness or rock type as a result of differences in the sedimentation rate or environment of deposition (Figure 7.9). In these cases, a bed of rock may be in the same position in a stratigraphic sequence, but it is somewhat different in thickness or rock type. A lateral change in rock type may be caused by a lateral change in depositional environment. For example, river channel deposits pass laterally into floodplain or delta deposits. Similarly, beach sands pass laterally into deeper water silts, muds, and clays.

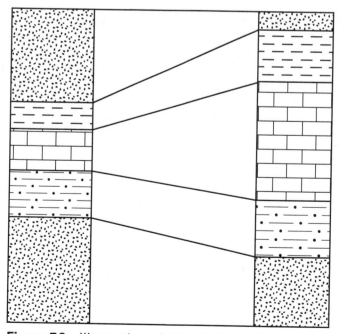

Figure 7.9   Illustration of lateral change in bed thickness.

In some cases, a bed thins progressively in one direction until it pinches out (Figure 7.10). A **pinch-out** might or might not be accompanied by the increase in thickness of an adjacent unit. In some case, the entire sedimentary section thins in a certain direction.

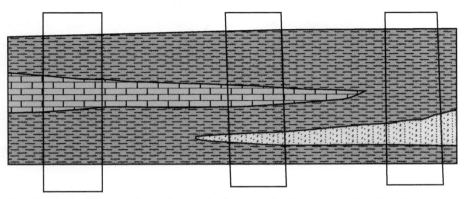

Figure 7.10 Illustration of the pinch-out of a limestone bed and a sandstone bed. Three stratigraphic sections are superimposed on the diagram, highlighting the difference in the sequence of facies present in different areas. Note that the pinch-outs produce a V-shaped pattern.

Walther's Law and Vertical Facies Changes

The sedimentary sequence exposed in outcrops is the result of different types of sediment being deposited in different sedimentary environments over time, producing a vertical sequence of different facies.

Lateral changes in facies are relatively easy to understand. Vertical facies changes can initially be somewhat puzzling. How does one layer of sedimentary rock come to overlie another? The vertical relationships between facies are explained by changes in sea level or changes in subsidence and sedimentation rates.

As laterally adjacent sedimentary environments shift back and forth through time, as a result of sea level change, facies boundaries also shift back and forth. Given enough time, *facies that were once laterally adjacent will shift so that the deposits of one environment come to overlie those of an adjacent environment*. In fact, this is how many (if not most) vertical sequences of sedimentary rocks were formed. This concept was first stated by Johannes Walther in 1894, and is called **Walther's law**. Basically, in a conformable sedimentary sequence (i.e., one without unconformities), sedimentary units that lie in vertical succession represent the deposits of laterally adjacent sedimentary environments migrating over one another through time.

At any one time, sediment of different types is being deposited in different places (Figure 7.11). Sand is deposited on the beach, silt is deposited offshore, clay is deposited in deeper water, and carbonate sediment is deposited far from shore (or where there is little or no input of terrigenous sediment). Sedimentary environments (and facies) move as sea level changes or as a basin fills with sediment.

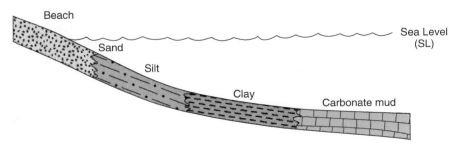

Figure 7.11 Distribution of sedimentary facies.

A sea level *rise* is called a **transgression**. A transgression produces a vertical sequence of facies representing progressively *deeper* water environments (a deepening-upward sequence) (Figure 7.12). As a result, a transgressive sequence has finer-grained facies overlying coarser-grained facies (fining-upward from sand at the bottom, and then to silt, and then to shale, and perhaps to limestone). Transgressions can be caused by melting of polar ice caps, displacement of ocean water by increased sea floor spreading and undersea volcanism, or by localized sinking or subsidence of the land in coastal areas.

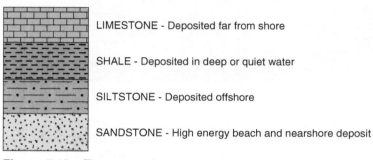

LIMESTONE - Deposited far from shore

SHALE - Deposited in deep or quiet water

SILTSTONE - Deposited offshore

SANDSTONE - High energy beach and nearshore deposit

Figure 7.12   Transgressive sequence.

A sea level *drop* is called a **regression**. A regression produces a sequence of facies representing progressively *shallower* water environments (a shallowing-upward sequence) (Figure 7.13). As a result, a regressive sequence has coarser-grained facies overlying finer-grained facies (coarsening-upward). Regression can be caused by a buildup of ice in the polar ice caps or by localized uplift of the land in coastal areas.

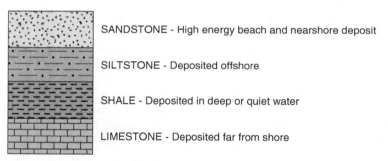

SANDSTONE - High energy beach and nearshore deposit

SILTSTONE - Deposited offshore

SHALE - Deposited in deep or quiet water

LIMESTONE - Deposited far from shore

Figure 7.13   Regressive sequence.

We can easily see how transgressive and regressive sequences form. These sequences are illustrated in Figure 7.14; *transgressive* sequences are shown in Figure 7.14A through D, and *regressive* sequences are shown in Figure 7.14E through G.

Figure 7.15 illustrates a transgression followed by a regression. We call this a **transgressive–regressive sequence**. The parts of the record deposited during the transgression, sea level high stand, and regression are labeled in the box to the right of the diagram. Four facies are shown: a sandstone facies, a siltstone facies, a shale facies, and a limestone facies. Note that the facies pattern produces a *broad V shape* in cross section.

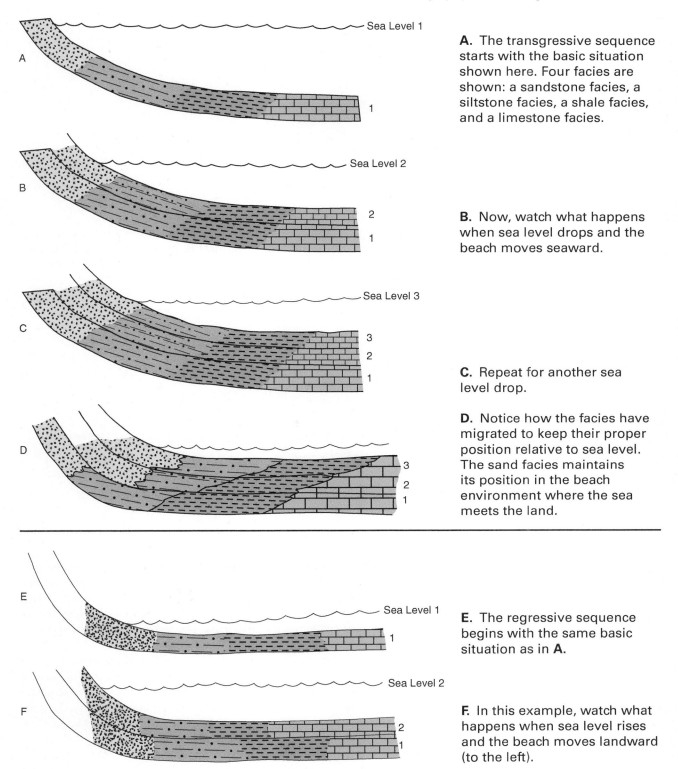

A. The transgressive sequence starts with the basic situation shown here. Four facies are shown: a sandstone facies, a siltstone facies, a shale facies, and a limestone facies.

B. Now, watch what happens when sea level drops and the beach moves seaward.

C. Repeat for another sea level drop.

D. Notice how the facies have migrated to keep their proper position relative to sea level. The sand facies maintains its position in the beach environment where the sea meets the land.

E. The regressive sequence begins with the same basic situation as in A.

F. In this example, watch what happens when sea level rises and the beach moves landward (to the left).

Figure 7.14 The building of transgressive and regressive sequences *(Continued).*

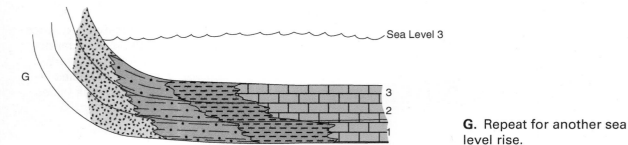

G. Repeat for another sea level rise.

Figure 7.14 The building of transgressive and regressive sequences.

Notice again how the facies have migrated to keep their proper position relative to sea level. The sand facies maintains its position in the beach environment, where the sea meets the land. The facies are marching inland as sea level rises.

What would happen next if sea level began to fall?

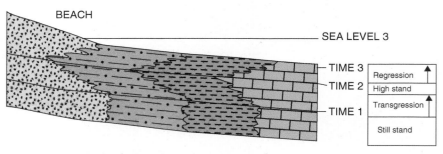

Figure 7.15 Transgressive–regressive sequence. The rock units in this diagram are diachronous.

Three **time lines** are shown (labeled Time 1, Time 2, and Time 3). (In the geologic record, a time line could be marked by a thin volcanic ash bed, representing a particular volcanic eruption event, or by the presence of a particular fossil that lived at a specific time with a wide geographic range, called an **index fossil**.) Note that the facies or lithologic units cut across the time lines. The facies are time-transgressive or **diachronous**.

Look carefully at the time line marked "Time 2," and follow it across the diagram. Time line 2 bisects the V shape of the transgressive–regressive sequence. The point of sea level high stand (maximum transgression) in a transgressive–regressive sequence is always a time line marking the time of maximum transgression. Similarly, the point of sea level low stand (maximum regression) in a transgressive–regressive sequence is always a time line marking the time of maximum regression.

Now let's see how a transgressive–regressive sequence looks in the stratigraphic record. Examine the three stratigraphic sections in Figure 7.16. Note that the sequence of facies in each section is different. How do these three sections relate to what you learned about transgressive-regressive sequences? Hint: Look at Figure 7.15.

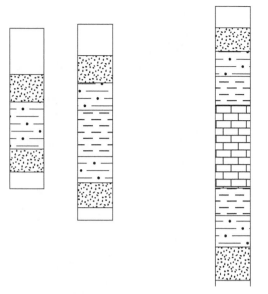

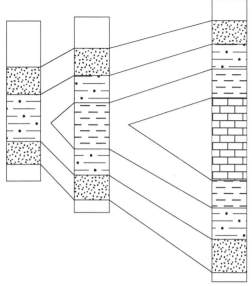

A. Three stratigraphic sections. Note that the sequence of facies in each section is different.

B. Draw lines between these three stratigraphic sections to correlate the facies.

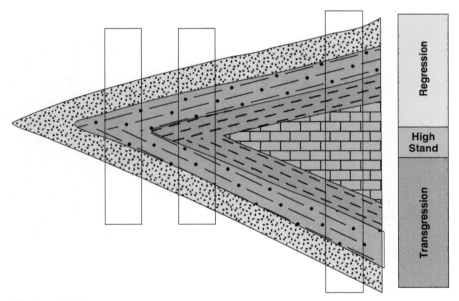

C. Interpret transgression, regression, and sea level high stand, and sketch in the facies between the sections.

Figure 7.16

Figure 7.16C shows the V-shaped pattern produced by migrating facies during a transgression followed by a regression. Three stratigraphic sections are superimposed on the pattern to illustrate how the facies would appear in vertical section in three different locations. Note that the facies present in each section are different as a result of the pinch-out. Draw a dashed line along the center of the V connecting the maximum landward extent of each facies. This is a **time line** marking the time of sea level high stand or maximum transgression.

A Real-Life Example of a Transgressive–Regressive Sequence, Dakota Hogback, Colorado

The rocks of the Dakota Hogback were originally deposited as flat layers of sediment during the Cretaceous Period, and they were tilted during the uplift of the Rocky Mountains (Figure 7.17).

In the photo on the right, you can see a thick dark gray rock unit overlain and underlain by tan rock units. This is a transgressive–regressive sequence (Box 7.2). The dark gray unit was deposited in a marine environment when sea level was high. The tan rock units above and below it were mainly deposited in fluvial environments. The reddish brown rocks below, on the left, were also deposited in nonmarine (or continental) environments.

The lower part of the rock sequence, from the reddish brown and tan rocks to the gray rocks, represents a transition from nonmarine to marine environments. This indicates that a **transgression** (or sea level rise) occurred. The upper part of the rock sequence represents a transition from the dark gray marine rocks back into the tan nonmarine rocks. This indicates that a **regression** (or drop in sea level) occurred. Taken together, the entire sequence of rocks documents the flooding of this area by the Cretaceous epicontinental seaway about 100 million years ago, followed by a drop in sea level.

Can you spot the **unconformity** in the photo on the left? The lower tan sandstone (on the left) varies in thickness across the outcrop. It is thinner high on the

Pamela Gore

Pamela Gore

Figure 7.17 Transgressive–regressive sequence exposed west of Denver, Colorado, along Highway 285 through the Dakota Hogback at Turkey Creek.

Box 7.2 Transgressive–Regressive Sequence through the Dakota Hogback

- This sequence of units (red nonmarine siltstones overlain by tan fluvial sandstone, overlain by dark gray marine shale) represents a transgression.
- The transgression was followed by a regression.
- The complete sequence of rocks, shown in Figure 7.17, includes both a transgression and a regression. We call this a transgressive–regressive sequence.

cliff, and it thickens toward the road level. On the picture on the left, use your finger or a pencil to follow the lower contact of the tan sandstone where it touches the underlying reddish brown rocks. This contact is a type of unconformity called a **disconformity**.

In Figure 7.17, the greenish-gray lowermost unit (on the left) belongs to the Upper Jurassic Morrison Formation and consists of siltstones, mudstones, and claystones with freshwater fossils deposited in nonmarine, continental sedimentary environments, such as lacustrine (lake) and fluvial environments. In some places, dinosaur bones and tracks are present in the Morrison Formation.

The Lytle Formation is dominated by reddish brown mudstones, sandstones and conglomerates that were mostly deposited in fluvial environments. The tan and gray rocks overlying the Lytle Formation belong to the Lower Cretaceous Dakota Group. The lower tan sandstone unit is the Plainview Sandstone Member of the South Platte Formation of the Dakota Group. The Plainview Sandstone unconformably overlies the Lytle Formation. The Plainview Sandstone is interpreted to have been deposited in a coastal plain swamp, tidal flat and fluvial environment.

The thick dark gray shale is called the Skull Creek Shale, and it was deposited in a marine environment.

An upper unit of tan Dakota Group sandstone overlies the dark gray shale on the far right. This unit is the Lower Cretaceous Horsetooth Member of the Muddy ("J") Sandstone of the South Platte Formation of the Dakota Group. This unit contains several uranium bodies and also serves as an important oil and gas reservoir. This sandstone was mostly deposited in fluvial environments, with some nearshore marine and deltaic deposits. This sandstone unit was deposited during a regression of the Cretaceous epicontinental seaway around 97 to 99 million years ago.

If we were to draw a stratigraphic section for this outcrop, it might look like Figure 7.18.

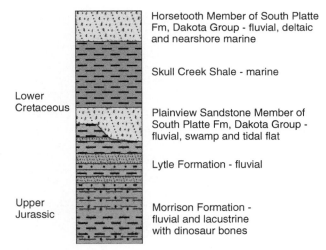

Figure 7.18 Stratigraphic section of the transgressive–regressive sequence through the Dakota Hogback exposed west of Denver, Colorado, along Highway 285 at Turkey Creek.

SELECTED REFERENCES

Boyd T. Geologic Overview of Jefferson County, Colorado. http://www.mines.edu/fs_home/tboyd/Coal/geology/overview.html

Higley DK, Cox DO. Oil and Gas Exploration and Development along the Front Range in the Denver Basin of Colorado, Nebraska, and Wyoming. In Higley DK, editor. Petroleum Systems and Assessment of Undiscovered Oil and Gas in the Denver Basin Province, Colorado, Kansas, Nebraska, South Dakota, and Wyoming—USGS Province 39. Reston, Va., U.S. Geological Survey, 2007. http://pubs.usgs.gov/dds/dds-069/dds-069-p/REPORTS/69_P_CH_2.pdf.

Keller J, U.S. Geological Survey. Mineral Resources of Jefferson County, Colorado, 2001. http://geosurvey.state.co.us/education/Documents/MineralResourcesofJeffersonCounty.pdf

Kellogg K, Klein T. Central Colorado Assessment Project. http://minerals.cr.usgs.gov/projects/colorado_assessment/

McCreary J. Colorado Geology Photojournals. http://www.cliffshade.com/colorado/dakota_hogback/

Stratigraphy and Lithologic Correlation Exercises

PRE-LAB EXERCISES

Dakota Hogback Stratigraphic Section

For the Dakota Hogback stratigraphic section in the figure below, answer the following questions.

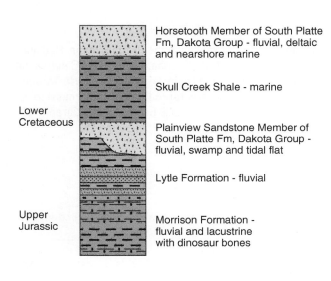

Horsetooth Member of South Platte Fm, Dakota Group - fluvial, deltaic and nearshore marine

Skull Creek Shale - marine

Plainview Sandstone Member of South Platte Fm, Dakota Group - fluvial, swamp and tidal flat

Lytle Formation - fluvial

Lower Cretaceous

Morrison Formation - fluvial and lacustrine with dinosaur bones

Upper Jurassic

1. List three rock types in the fluvial Lytle Formation, as shown by the lithologic symbols.

   _____

   _____

   _____

2. Label the position of the unconformity.

3. What type of unconformity is this? _____

4. Label the part of the section that represents sea level high stand.

5. Label the part of the section that represents a transgression.

6. Label the part of the section that represents a regression.

Cambrian Strata in the Grand Canyon

For the diagram below showing Cambrian strata in the Grand Canyon area, answer the following questions.

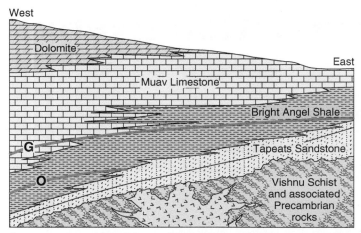

Cross-section of Cambrian sedimentary rocks exposed in the Grand Canyon. Red line O marks the top of the Low Cambrian *Olenellus* trilobite zone, and red line G marks the Middle Cambrian *Glossopleura* trilobite zone. (From Levin, H., 2013, *The Earth Through Time* (10th edition), figure 10-12, p. 285. This material is reproduced with permission of John Wiley & Sons, Inc.). Adapted from Mckee, E. D., 1945, Cambrian Stratigraphy of the Grand Canyon Region, Washington, Carnegie Institute, Publication 563.

1. Label the position of the unconformity.

2. What type of unconformity is this? _____

3. What does the vertical sequence of rock types indicate about sea level changes? Is this a transgressive sequence or a regressive sequence? _____

4. What is the most likely depositional environment of the Tapeats Sandstone in this sequence? _____

5. What is the most likely depositional environment of the Bright Angel Shale in this sequence? _____

   Notice the two red lines, one labeled O and the other labeled G. These lines indicate the times at which two different trilobite species (*Olenellus* and *Glossopleura*) lived, along with their geographic distribution. These trilobites are index fossils or indicators of very specific intervals of geologic time. *Olenellus* lived during the Early Cambrian and *Glossopleura* lived during the Middle Cambrian. These red lines can be considered time lines.

6. What is the age of the Bright Angel Shale in the western part of this diagram?

   _____

7. What is the age of the Bright Angel Shale in the eastern part of this diagram?

   _____

8. By examining these trilobite distributions, we can see that the Bright Angel Shale is diachronous (it is older in some areas and younger in others). Is the Bright Angel Shale older in the east or older in the west? _____

9. As sea level changed, during the deposition of the Tapeats Sandstone (and overlying rock units), was the sea mainly to the east of this area or to the west of this area? _____

Lithologic Correlation

In this next part, you will practice lithologic correlation. You need to use the following:

- Ruler
- Pencil with eraser (no pen)
- Colored pencils (optional)

To correlate the sections, you will use a pencil and a ruler to draw lines connecting geologic contacts between beds as illustrated in Figure 7.16. This is not a matching activity, so do not draw lines to the midpoints of the units. *Draw your lines at the contacts*. Use a ruler and be as neat as possible.

Note that each vertical column is a **stratigraphic section**. Each different rock type can be regarded as a bed or facies.

1. Correlate the three stratigraphic sections below. Draw lines between the three stratigraphic sections below to connect the geologic contacts between beds.

a. How many beds can be correlated across all three sections?_____

b. How thick is the thickest stratigraphic section (entire column of rocks)? (Note that the scale goes from 0 to 70 meters.) _____ meters

c. A bed of coal (black) is present in sections B and C. Draw where it would appear in the blank area of section A. How deep would you have to drill in section A (starting at the top of the section, up near the letter A) to reach the buried coal seam? _____ meters

d. What type of unconformity is present? _____

e. Above the unconformity, is there a transgressive sequence or a regressive sequence? _____

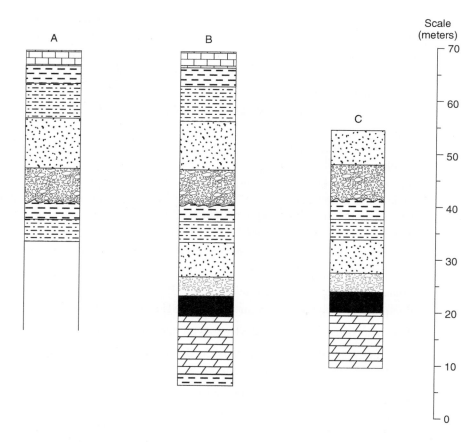

2. Correlate the two stratigraphic sections below, and answer the questions. Draw lines to correlate the two sections below. Use a ruler to connect the contacts between rock types.

   a. Which section (A or B) contains an unconformity? _____

   b. Label the position of the unconformity with an arrow and the word "unconformity". _____

   c. What type of unconformity is it? _____

   d. Which rock types are missing from that section because of the unconformity?

   _____   _____

Scale
(meters)

A

B

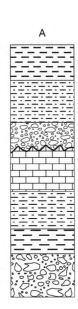

4.0

3.0

2.0

1.0

0

3. List the rock types in section A, above, in order from youngest to oldest.

_____

_____

_____

_____

_____

_____

_____

LAB EXERCISES

1. The photo below was taken near Menlo, Georgia, along Route 48. There is a thick resistant layer of conglomeratic sandstone bed overlying silty shale, with a thin coal seam (less than an inch thick) at the contact between the two.

Pamela Gore

Sandstone overlying coal and shale, northwestern Georgia.

    a. In the box beside the photo, draw a simple stratigraphic section to correspond to the rock outcrop in the photo. Be sure to use the proper lithologic symbols. Fill the entire box with beds and lithologic symbols.

    b. Considering that you have a coal deposit between the sand and the shale, which sedimentary environments are likely to be represented here?

    _____

    c. If you were in the field with the people in the photo, what would you look for to distinguish whether these rocks were marine deposits or nonmarine deposits?

    _____

2. The photo below was taken near Hancock, Maryland, along Interstate 68 at Sideling Hill.

Pamela Gore

Sideling Hill Road Cut, Interstate 68, Hancock, Maryland

Here is some information about the 340-foot high Sideling Hill road cut:

- The dark gray to black unit near the top of the hill is coal and black shale.
- The light-colored unit above the coal and black shale is interbedded conglomerate and sandstone.
- The light-gray to white unit below the coal and black shale is sandstone.
- The pinkish to reddish brown units below the sandstone are interbedded sandstone and shale. The darker beds are shale and the lighter beds are sandstone.
- All of the above units belong to the Purslane Formation.
- The dark maroon to purplish brown units on the right side of the road cut are interbedded siltstone (darker color) and sandstone (lighter color). There is at least one thin bed of dark gray to black coal in this part of the road cut, and there are a few thin beds of shale.
- These beds in the lower part of the section belong to the Rockwell Formation.
- Both the Purslane Formation and the Rockwell Formation are Mississippian in age.

  a. What type of fold is shown in the photo above?

  _____

  b. For the same photo, draw a stratigraphic section in the box to the right of the photo, using the proper lithologic symbols, and estimating the proper position of each unit from its thickness in the photo. Fill the entire box with beds and lithologic symbols. (Suggestion: Lay a piece of paper across part of the image, with the edge of the paper between the top of the hill and the lower right corner of the image. Draw lines at the contacts between the major rock units. Use the information in the box below the photo to help you fill in the rock types. Then transfer this information onto your stratigraphic section.)

    c. In which sedimentary environment(s) were the dark gray to black sedimentary rocks of the Purslane Formation most likely deposited? _____ _____

    d. In which sedimentary environment(s) were the lighter-colored units of the Purslane Formation most likely deposited? _____ _____

    e. In which sedimentary environment(s) were sedimentary rocks of the Rockwell Formation most likely deposited? _____

3. Correlate the three stratigraphic sections below and answer the questions.

    a. Use a ruler and pencil draw lines to correlate the three stratigraphic sections. Connect the contacts between the beds. Note that the units differ in thickness in each section.

    b. What happens to the conglomerate in section C? Why is it not present in section A or B? _____

    c. Why does the thickness of the sandstone change from section A to section B? _____

    d. In which direction (west or east) was the land? _____

    e. In which direction (west or east) was deeper water? _____

    f. Label the right side of section C to illustrate a transgression, a regression, and the time of sea level low stand.

    g. Label the pinch-out with an arrow and the term "pinch-out."

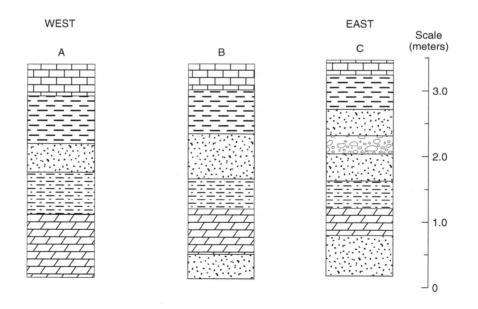

4. Correlate the three stratigraphic sections below and answer the questions.

   a. Use a ruler to draw lines to correlate the three stratigraphic sections. Connect the contacts between the beds. Note that some of the units differ in thickness in each section.

   b. Label the left side of section A to illustrate a transgression, a regression, and the time of sea level high stand.

   c. These sediments were deposited in an ancient sea. Based on the rock types, which facies (i.e., rock type) was deposited nearest to the mainland?

   _____

   d. Which facies (i.e., rock type) was deposited farthest from the land?

   _____

   e. Why is the limestone absent from section C? _____

   f. In which direction (east or west) was the land? _____

   g. Explain your reasoning for your answer to question f.

   _____

   _____

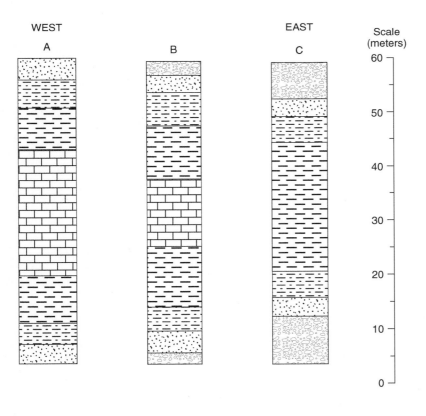

5. The following four sections are based roughly on drill cores from four wells.

    a. Draw correlation lines between the four sections below. Look for unique marker beds to help you.

    b. In which sedimentary environments were these rocks most likely deposited?

      _____

      _____

    c. What type of unconformity is present? _____

    d. Can you spot any lateral facies changes? Describe them.

      _____

      _____

    e. What does the presence of gypsum indicate about the climate at the time of deposition? _____

    f. What does the presence of coal and black shale with fish fossils indicate about the climate at the time of deposition?

      _____

    g. Can you pick out any evidence of changing climate through time? Explain.

      _____

      _____

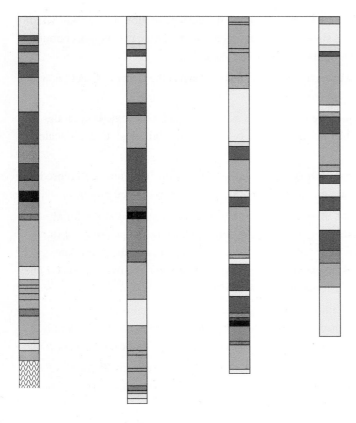

Key

Conglomerate

Red sandstone

Red shale

Fining-upward sequences of brown or
red sandstone, siltstone, and shale

Gray sandstone, siltstone, and shale

Coal

Black shale with freshwater fish fossils

Limestone

Gypsum

Metamorphic rocks

6. Using the data from the western and eastern sections in the boxes below, do the
   following:

   a. Using a pencil and a ruler, draw two stratigraphic sections on graph paper using
      the rock type and thickness data provided. You may use your own graph paper or
      the printed graph paper included here. Draw a horizontal line across the top of your
      paper to represent the ground surface. Draw the sections below this line as if they
      were drill core data (plotting each unit below the ground surface).

Draw the western section on the left and the eastern section on the right. Each stratigraphic section should be about an inch wide. Draw the sections about 3 to 4 inches apart, with their tops on the same line.

The first unit listed in each section should be drawn at the top of that section, with all of the other units in order below.

Be sure that you plot the correct thicknesses for each lithologic unit on the graph paper. All thicknesses must be to scale. You might wish to use a scale of 1 inch = 100 feet.

Draw the appropriate lithologic symbols on each unit. You may use colored pencils. You may use a computer to draw your sections if you prefer.

b. Locate the positions of all unconformities in the two sections (look for abrupt or erosional contacts in the section descriptions). Mark the positions of the unconformities on your sections by using a wavy line, and then correlate the unconformities by extending the wavy line between the two sections. Label your wavy lines "Unconformity" for clarity.

c. Correlate the two sections by drawing lines to connect the contacts between equivalent units. Remember that some beds might change in thickness laterally, whereas others might pinch out or be eroded away in one of the sections.

d. Label each section to illustrate all transgressions (T), regressions (R), sea level high stands, and sea level low stands. Observe carefully. You will have to think about the sedimentary environments to identify sea level changes.

WESTERN SECTION	
100 ft	Shale; dark gray, fissile; rare ammonoid fossils; lower contact abrupt, probably erosional.
40 ft	Basalt with vesicles (lava flow).
35 ft	Conglomerate; red to brown; no fossils.
100 ft	Sandstone; white to pale red; dominantly quartz sandstone, very well-sorted, rounded sand grains, cross bedded; no fossils.
75 ft	Dolostone; tan to light gray; a few gastropod fossils present.
40 ft	Limestone; light gray; oolitic limestone.
25 ft	Limestone; medium gray; fossiliferous limestone; tabulate corals, crinoids, and bryozoans abundant. Lower contact abrupt.
50 ft	Conglomerate and breccia; brown to red matrix; no fossils.
75 ft	Sandstone with some conglomerate beds; brown to red; plant fossils and rare bones present.
100 ft	Siltstone (brown to green) with rare coal beds; plant fossils and some bivalves.
50 ft	Shale; green to gray or brown; trilobites and brachiopods present.
690 ft	TOTAL

EASTERN SECTION	
100 ft	Shale; dark gray; fissile; rare ammonoid fossils present; lower contact abrupt - probably erosional.
40 ft	Basalt with vesicles (lava flow).
75 ft	Sandstone; white; dominantly quartz sandstone, very well-sorted, rounded sand grains; cross bedding present; fossils absent.
100 ft	Dolostone; tan to gray; a few gastropod fossils present.
60 ft	Limestone; light gray; cross bedded, oolitic limestone.
25 ft	Limestone; medium gray; fossiliferous limestone with abundant tabulate corals.
20 ft	Limestone; dark gray micrite limestone; lower contact abrupt, apparently erosional.
65 ft	Sandstone with local conglomerate beds; brown to red.
50 ft	Siltstone (brown, green, and gray), with rare coal beds; plant fossils present.
40 ft	Shale; gray to brown; fissile; trilobite fossils present.
100 ft	Limestone; gray; micrite limestone; rare brachiopods and bryozoans.
675 ft	TOTAL

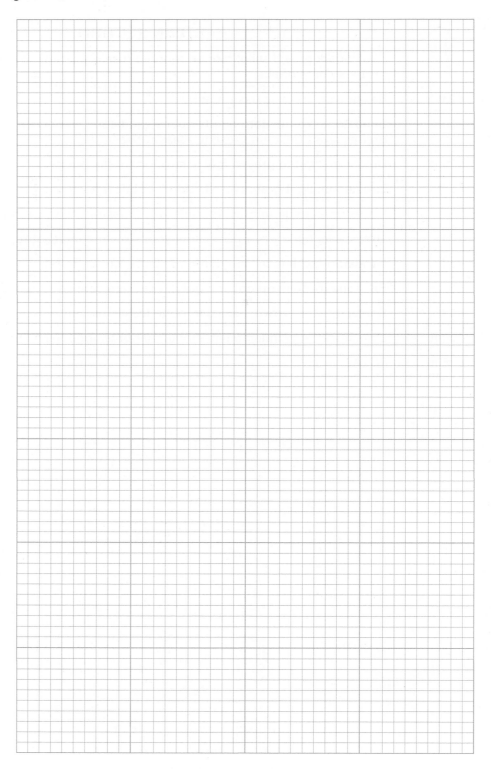

OPTIONAL ACTIVITY

Take a field trip to an outcrop or road cut to observe a sequence of sedimentary rocks and measure a stratigraphic section. You will need a tape measure or meter stick and a notebook and pencil, and you need to be able to identify the sedimentary rocks.

Fossils on the Internet

In this lab, you will visit several websites to learn about fossils and paleontology. You will examine fossils at the University of California Berkeley Museum of Paleontology website. You will also learn about what paleontologists do and about careers in the geosciences. The lab consists entirely of research using online resources.

GETTING INTO THE FOSSIL RECORD

http://www.ucmp.berkeley.edu/education/explorations/tours/fossil/index.html

Click on student level 2 on the right side of the page. Then, click on the links at the bottom of the page (MORE and NEXT) to continue through the site. Some pages require you to click on an image or answer a question before the link appears to guide you to the next page.

1. View the animation about getting into the fossil record, and write several sentences summarizing how a dinosaur can become a fossil.

2. The word *fossil* (from Latin) means _____

3. What are paleontologists? _____

4. Fossils can be body parts of ancient organisms, or they can be traces. Give five examples of traces.

   _____
   _____
   _____
   _____
   _____

5. Which type of organism do you think is most likely to be preserved? One that gets buried quickly or one that gets buried slowly?

   _____

6. What are three biotic factors that can affect an organism after death?

   _____
   _____
   _____

7. What is an abiotic factor that can prevent the organism from becoming preserved *after* it has been buried?

   _____

8. What is a mold? _____

9. What is a cast? _____

10. What is amber? _____

11. What type of animal might become preserved in amber?

    _____

12. Where are the Rancho La Brea Tar Pits? _____

13. What type of animal might become preserved in a tar pit?

    _____

14. Which is more likely to fossilize: hard parts or soft parts?

    _____

15. Of all the organisms alive today, what percentage is most likely to be preserved as fossils?

    _____

16. What are three ways a fossil can be destroyed after it has formed?

    _____

    _____

    _____

17. Which type of rock is most likely to contain fossils? Igneous, sedimentary, or metamorphic?

    _____

18. Why is a mammoth more likely to fossilize than a caterpillar?

    _____

STORIES FROM THE FOSSIL RECORD

http://www.ucmp.berkeley.edu/education/explorations/tours/stories/middle/intro.html

Note the four sections on this page: Paleoecology, Past Lives, Geologic Time and Biodiversity. Click on the links at the bottom of the page (MORE and NEXT) to continue through the site. Some pages require you to click on an image or answer a question before the link appears to guide you to the next page. After completing each section, return to this page to start the next section.

Paleoecology

1. What is paleoecology?

   _____

2. What are three abiotic factors of the ecosystem?

_____

_____

_____

3. What are Archaeocyathids?

_____

4. What could explain finding fossils of sponges in the Nevada desert?

_____

5. What does _terrestrial_ mean?

_____

6. Plants with leaves that have smooth edges grow in what sorts of climates?

_____

7. In what climate do plants with leaves with toothed edges live?

_____

8. List two common interactions between organisms.

_____

_____

9. What are the holes in the ammonite shell? _____

10. What animal made the holes in the ammonite shell? _____

11. The taiga forest ecosystem is composed mostly of which two types of trees?

_____

_____

12. What can fossil pollen and spores tell us? _____

Past Lives

1. Name a fossil animal with growth rings (from the website).

_____

2. How many species of trilobites existed? _____

3. Which fossil animal is sometimes found curled up (from the website)?

_____

4. Why did they curl up? _____

5. What is the evidence that the maiasaurs cared for their young in the nest?

_____

6. What is the evidence that hadrosaurs lived in herds with social behavior?

_____

7. In the page on the whale forelimb, what happened to the *rear* legs of the whale ancestor? _____

Geologic Time

1. What is superposition? _____

2. What is an index fossil? _____

3. What is *Turritella*? _____

4. What is the genus of the plant fossil that has been used to piece together the positions of the continents in the past? _____

5. When and where did *Glossopteris* live? _____

6. On what supercontinent did *Glossopteris* live? _____

Biodiversity

1. What is biodiversity? _____

2. What is the genus of the fossil bird? _____

3. Did *Archaeopteryx* have feathers? _____

4. Did *Archaeopteryx* have teeth? _____

5. What appears to be the closest living relative of the dinosaur?

_____

6. When did trilobites become extinct? _____

7. What is one of the closest living relatives of the trilobite?

_____

8. What percentage of all species that lived on Earth are now extinct?

_____

9. A mass extinction occurred about 248 million years ago. This was at the end of what geologic period?

_____

10. What are four factors contributing to the extinction?

_____   _____

_____   _____

11. What percentage of the animals went extinct 248 million years ago?

_____

12. List six groups of organisms that became extinct about 248 million years ago.

_____   _____

_____   _____

_____   _____

13. List four groups of organisms that became extinct about 65 million years ago.

_____   _____

_____   _____

14. What sorts of animals became extinct about 11,000 years ago? (List four types of mammals.)

_____   _____

_____   _____

15. What was the most likely cause of the extinction 11,000 years ago? (List two factors.)

_____

_____

16. When reef-building organisms went extinct, what became of the other organisms that inhabited the reef?

_____

17. What appears to be the closest living relative of the eurypterid?

_____

UNIVERSITY OF CALIFORNIA BERKELEY MUSEUM OF PALEONTOLOGY: HISTORY OF LIFE THROUGH TIME

http://www.ucmp.berkeley.edu/exhibits/historyoflife.php

1. What are the three _domains_ of life?

_____

_____

_____

Click on the link for BACTERIA in the diagram.

2. What do bacteria do, and why are they important? _____

_____

3. What is the age of the oldest fossil bacteria? _____

At the bottom of the Bacteria page, click on FOSSIL RECORD and follow the links.

4. What are cyanobacteria? _____

_____

5. Name two layered structures that cyanobacteria form.

_____

_____

6. Explain how cyanobacteria produce these layered structures.

_____

Use the BACK button to return to the "History of Life Through Time" page. Click on the link for EUKARYOTA in the diagram.

7. Which four groups of organisms are included in the Eukaryota?

_____

_____

_____

_____

Click on the PALM TREE.

8. Plants first appeared in the _____ but did not begin to resemble modern land plants until the Late _____.

9. Trees appeared by the end of the _____ Period.

Use the BACK button to return to the "History of Life Through Time" page. Click on the link for ARCHAEA in the diagram.

10. When were the Archaea discovered? _____

_____

11. What sorts of extreme environments do the Archaea inhabit? List five.

_____

_____

_____

_____

_____

Use the BACK button to return to the "History of Life Through Time" page. Click on LEARN MORE ABOUT PHYLOGENY AND CLADISTICS, or use this link: http://evolution.berkeley.edu/evolibrary/article/phylogenetics_01. Click on the NEXT button in the lower right corner of the page to proceed through the site.

12. What is phylogenetic systematics?

_____

_____

13. What is a clade? _____

14. What are the two main advantages of phylogenetic classification?

_____

_____

15. Why are Linnaean classification ranks (such as kingdoms, phyla, classes, and orders) misleading?

_____

16. How do biologists deal with phylogenetic classification? Can you still use Linnaean names? _____

17. What are the three basic assumptions in cladistics?

_____

_____

_____

18. A phylogeny is a hypothesis. How could a phylogeny change?

_____

19. Biologists use molecular characters to reconstruct relationships between lineages of organisms. What is meant here by _molecular?_ (Click on the word to see the definition.)

_____

20. What is parsimony? _____

21. Whales are more closely related to what sorts of mammals? _____

22. What feature did the whales have in common with modern hoofed mammals like hippos, camels, cows, and giraffes? _____

23. How can knowledge of evolutionary trees be used to develop new drugs to fight cancer? _____

24. Give an example: _____

GEOLOGIC TIME

University of California Berkeley Museum of Paleontology: Tour of Geologic Time

http://www.ucmp.berkeley.edu/exhibit/histgeoscale.php

1. Explain the contributions of Nicholaus Steno to geology.

_____

_____

2. What were two of William Smith's contributions to geology? (What did he produce, _and_ what was the principle for which he is responsible?)

_____

_____

3. The beginning of the Phanerozoic is marked by what occurrence? (Read the whole paragraph carefully.)

_____

4. What are the three divisions (Eras) of the Phanerozoic Eon?

_____

_____

_____

5. What do the following terms mean?

-zoic _____

Cen- _____

Meso- _____

Paleo- _____

6. What major group of animals dominated the Mesozoic Era? _____

FREQUENTLY ASKED QUESTIONS ABOUT PALEONTOLOGY

http://www.ucmp.berkeley.edu/FAQ/faq.html

1. What is paleontology? _____

2. What are the practical uses of paleontology? (Give at least three.)

_____

_____

_____

3. How do paleontologists know how old fossils are?

_____

_____

_____

4. Fossils that are most useful for correlation tend to be: (List four things.)

_____

_____

_____

_____

5. How can you find fossils in your area?

_____

_____

_____

6. What organizations exist for paleontologists in your area? (See also
   http://paleo.cc/kpaleo/paleorgs.htm.)

_____

_____

_____

GEOSCIENCE CAREERS

Visit the U.S. Bureau of Labor Statistics to learn about careers in the Geosciences.
http://www.bls.gov/ooh/life-physical-and-social-science/geoscientists.htm#tab-1

Click on the link in the blue box on the lower right corner of the page to proceed through the site.

1. What do geoscientists do?

_____

_____

_____

2. What are some examples of types of geoscientists? (List three that you find interesting.)

_____

_____

_____

3. What are the top industries employing the largest number of geoscientists? (List four.)

_____

_____

_____

_____

4. Where do geoscientists work? (List three general places).

_____

_____

_____

5. What education do geoscientists need?

_____

6. What important qualities or skills do geoscientists need? (List three).

_____

_____

_____

7. What is the median annual wage of geoscientists?

_____

8. Which industry pays the highest salary, and what is that salary?

   _____

9. What is the employment outlook for geoscientists, according to the U.S. Bureau of Labor Statistics? Will employment of geoscientists grow faster or slower than average?

   _____

10. What is the best degree to have to increase your employment prospects?

   _____

11. How do gasoline prices relate to employment in the geosciences?

   _____

Microfossils and Introduction to the Tree of Life

WHAT ARE MICROFOSSILS?

Microfossils are fossils that are too small to be studied without the aid of a microscope. They include the remains of protists and small parts of multicellular plants and animals. **Protists** are mostly single-celled eukaryotic organisms that are neither animals nor plants but that can have characters of both, such as movement and the ability to photosynthesize. **Eukaryotic organisms** are those that have cells with a nucleus and organelles.

In this lab, you will learn about several groups of microfossils that can be used to date rocks (through biostratigraphy) and to interpret sedimentary environments.

HOW ARE MICROFOSSILS USED IN GEOLOGY?

Microfossils are useful for determining the *age* of sedimentary rocks and for interpreting the environment where the sediments were deposited. Microfossils are often used for **biostratigraphic correlation**. This means that by studying the microfossils present in different outcrops or drill cores, it is possible to correlate the stratigraphic sections. Microfossils have been used for biostratigraphic correlation with great success in the petroleum industry. In drill cores hundreds or thousands of feet long, there may be many beds with the same rock type, but each bed has a unique assemblage of microfossil species. Thus, it is possible to correlate beds between cores or rock outcrops based on the assemblages of microfossils present in each bed.

The amount of time between the evolution (or first appearance) of a species and the extinction (or disappearance) of a species is called its **geologic range**. Many species of microorganisms have short geologic ranges (i.e., they lived during a very specific interval of time) and were widely distributed in the oceans. These microfossils serve as good **index fossils** or indicators of very specific intervals of geologic time, which can be used as tools for worldwide time correlation. These contrast with other organisms that have long geologic ranges and might have inhabited geographically restricted environments, such as reefs or tidal flats. These organisms are not useful for correlation, but they can serve as tools for interpreting ancient sedimentary environments.

Several characteristics make a microfossil a good **index fossil:**

1. **Short geologic range:** They only lived during a short span of geologic time.
2. **Widespread geographic distribution:** Many microscopic organisms are part of the floating plankton in the oceans. These species are readily distributed over large areas by ocean currents.

3. **Facies independence:** They are not restricted to any particular rock type; they are present in many rock types. For example, dead plankton would settle into any sediment that is accumulating on the sea floor.

4. **Distinctive and easily recognized form**

5. **Preservable, fossilizable hard parts**

6. **Abundance:** The fossils should be abundant enough to be collected in sufficient quantity for study. Because they are so tiny, hundreds or thousands of microfossils may be present in small sediment samples.

MICROFOSSILS AND THE TREE OF LIFE

The relatively recent advances in gene sequencing and molecular DNA analysis (molecular phylogenetics) have allowed scientists to compare selected genes or the entire genome of a wide variety of protists and multicellular organisms using high-powered computers. Molecular genetic studies tend to be supported by biochemical data and morphological data from ultrastructural research on cells (study of the details of cell structure with an electron microscope). As a result, evolutionary and structural relationships have been verified between certain groups of organisms, prompting scientists to offer a *complete revision of the classification of the eukaryotes.*

The terminology used for taxonomy or classification of organisms changed in 2005,[1] and some traditional groupings of organisms have been abandoned. Instead of kingdoms, phyla, classes, and so on, there are now nameless ranks, with the ranked hierarchy indicated by indented paragraphs or bullets. This approach has the advantage of flexibility and ease of modification with rapidly occurring advances in the study of gene sequencing. The traditional kingdoms of Animalia, Plantae, Fungi, and so on are now seen to be derived from lineages of single-celled organisms.

Domain terminology is used; a **domain** is a taxonomic rank higher than a kingdom. The term was introduced in 1990. There are three domains:

1. **Domain Bacteria** are single-celled organisms that lack a nucleus (prokaryotes).

2. **Domain Archaea** are single-celled microorganisms that lack a nucleus (prokaryotes) but evolved independently from bacteria, as shown by a different biochemistry.

3. **Domain Eukaryota** or **Eukarya** are organisms with cells that have a nucleus (or several nuclei) and other organelles. (Includes animals, plants, fungi, and protists).

The **Archaea** and **Bacteria** are prokaryotes because they lack nuclei; however, studies of genes and biochemistry suggest that Archaea are more closely related to the Eukaryota. Other aspects of their biochemistry are unique. Many Archaea inhabit extreme environments such as hot springs, geysers, black smokers, and oil wells, but they inhabit a wide variety of environments. We will not consider either of these groups in the lab other than to mention that stromatolites (an organosedimentary structure) are constructed through the activities of cyanobacteria (blue-green bacteria). Only the Eukaryotes are considered in the fossil labs in this course.

Eukaryotes (multicellular organisms and single-celled organisms with a nucleus) are now categorized into six supergroups on the basis of genetic similarities. The domains and supergroups can be diagramed as a tree of life (Figure 9.1). Several of these supergroups are represented by microfossils that we will study in this lab. Note: Keep in mind that the tree of life is currently in a state of flux, and these groupings will continue to change with further studies of molecular phylogenetics. Some subsequent studies have suggested grouping some of these, for instance, combining the Opisthokonta and the Amoebozoa into the Unikonts.

The six Eukaryote supergroups are:

1. **Amoebozoa:** Amoebas and slime molds (not.studied in this lab)

2. **Opisthokonta:** Animals (Metazoa), fungi, choanoflagellates (sponge relatives), and others

3. **Rhizaria:** Foraminifera, radiolaria, and others

4. **Archaeplastida** (or **Plantae**): Plants, green algae or charophytes, red algae, and others

5. **Chromalveolata:** Diatoms, dinoflagellates, coccoliths (golden-brown algae), silicoflagellates, and others

6. **Excavata:** Some flagellates (for example, *Euglena, Giardia, Trichomonas, Leishmania*) (not studied in this lab)

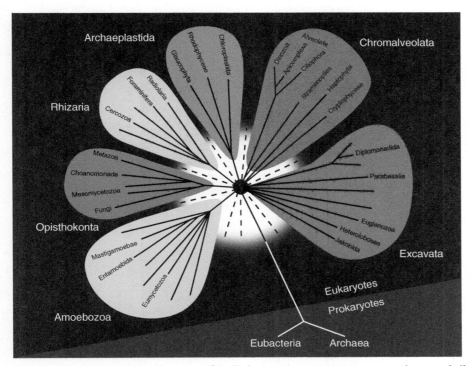

Figure 9.1 Tree of life diagram. Six Eukaryote supergroups are shown. Adl, S.M. et al., (2005), cover image of the *Journal of Eukaryotic Microbiology,* 52(5). Reprinted with permission from John Wiley & Sons, Ltd.

MICROFOSSIL GROUPS

Rhizaria

The **Rhizaria** are single-celled protists with pseudopods, some of which build tiny shells, called **tests**. The Rhizaria include **foraminifera** and **radiolaria**.

Most **foraminifera** (forams, for short) have skeletons of calcium carbonate, but some are organic walled, and some are agglutinated, (meaning that they are made of tiny bits of sand or shells that have been glued together) The shells (or tests) of forams accumulate on the sea floor to form limestone.

Radiolarians have distinctive porous shells made of silica and are common contributors to the formation of the sedimentary rock **chert**.

Foraminifera (Figure 9.2)

Geologic Range
Benthonic foraminifera or bottom dwellers: Cambrian to Recent (Quaternary).
Planktonic foraminifera or floaters: Jurassic to Recent (Quaternary).

Shell Composition
Calcite or aragonite. Some have shells with cemented grains (agglutinated). Some are organic.

Size
Most are 0.1 to 3.0 mm; some are larger, up to 1 cm or more.

Significance

Foraminifera are a source of carbonate sediment, are useful in biostratigraphy and marine paleoenvironmental interpretation, and provide paleotemperature determination using oxygen isotope ratios of their shells.

Morphology

May be coiled, straight, globular, or other in a wide range of shapes.

Environment

Marine only. There are benthic and planktonic foraminifera; the large ones are benthic.

A. Scanning electron microscope images of *foraminifera*.

Top row: *Elphidium*, *Nodosaria*, *Uvigerina*, and *Holmanella* (scale bar in each image is 100 μm).

Second row: *Cribroelphidium* (width of specimen is 250 μm)
Laevepeneroplis (width of specimen is about 250 μm)
Textularia (width of specimen is about 500 μm)
Nonion (width of specimen is about 250 μm)

B. Planktonic foraminifera viewed with a scanning electron microscope.

C. Foraminifera in shell hash, viewed through a binocular stereozoom microscope.

Figure 9.2 Foraminifera.

D. Large Lepidocyclinid foraminifera, Clinchfield Limestone, Georgia. (Penny for scale.)

E. *Nummulites* from the Eocene. The specimen has been split to show internal details. (Scale in millimeters.)

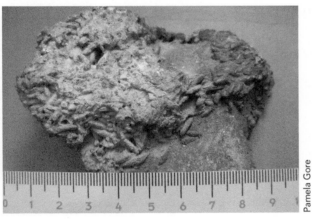

F. Large fusulinid foraminifera. (Scale in millimeters.)

Figure 9.2 Foraminifera (*Continue*).

Radiolaria (Figure 9.3)

Geologic Range
Cambrian to Recent (Quaternary).

Shell Composition
Silica (amorphous, opaline silica).

Size
0.1 to 2.0 mm.

Significance
Radiolarians are useful in biostratigraphy. They accumulate to form radiolarian ooze on the abyssal plain. They are also a constituent of chert.

Morphology
Microscopic spiny globes with large lacelike pores, or helmet shaped (or spaceship shaped) with large, lacelike pores. They are transparent and glassy.

Environment
Marine only; planktonic.

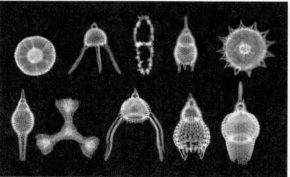

Image courtesy of The Wecskaop Project (What Every Citizen Should Know About Our Planet)

Silicious tests of ten species of marine radiolarians.

Figure 9.3   Radiolarians.

Chromalveolata: Photosynthetic Protists

The **Chromalveolata** are single-celled protists with plastids (like chloroplasts that contain pigments used in photosynthesis). The Chromalveolata include the diatoms, dinoflagellates, and coccolithophores, as well as another microfossil group not covered in this lab, the silicoflagellates.

The shells of **diatoms** are made of silica, and they accumulate to make up the rock diatomite. The shells of **coccolithophores** are called coccoliths. They are made of calcium carbonate and are the main constituent of chalk. Dinoflagellates are organic-walled.

Diatoms (Figure 9.4)

Geologic Range
Cretaceous to Recent (Quaternary).

Shell Composition
Silica.

Size
Most are 0.05 to 0.02 mm; some are up to 1 mm.

Significance
Diatoms are useful in biostratigraphy and paleoenvironmental interpretation. They are a major constituent of diatomite or diatomaceous earth, and they are an integral part of the food chain (phytoplankton). Diatoms are the most abundant phytoplankton in the modern ocean.

Morphology
Pillbox shape, consisting of two valves (shells) that may be circular, triangular, or elongate. Circular diatoms have radial ornamentation. Elongate diatoms have transverse markings. They are covered with pores.

Environment
Both marine and nonmarine or freshwater; planktonic or attached.

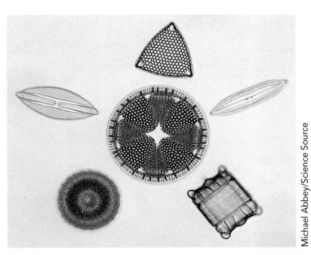

A. Modern marine diatoms with tests made of silica.

B. Fifty species of diatoms arranged in the shape of a circle.

Michael Abbey/Science Source

Image courtesy of The Wecskaop Project (*What Every Citizen Should Know About Our Planet*)

Figure 9.4 Diatoms

Dinoflagellates (Figure 9.5)

Geologic Range
Silurian to Recent (Quaternary).

Shell Composition
Organic material (sporopollenin).

Size
5 µm to 2 mm.

Significance
Dinoflagellates are useful in biostratigraphy and paleoenvironment interpretation. They are an integral part of the food chain (phytoplankton). Dinoflagellates cause red tides and secrete paralytic shellfish poison.

Morphology
Dinoflagellates are covered with a series of tiny plates and have an indentation around their equator that held a coiled flagellum in life. Shape is variable and can resemble a top or a star. Some are covered with spines.

Environment
Marine and nonmarine or freshwater; most are planktonic. Others are symbionts or parasites, including the zooxanthellae that live within the cells of corals.

Dinoflagellates are an important part of the food chain, and they can reproduce rapidly in what is called an **algal bloom**, resulting in a **red tide** (or red coloration of the seawater because of their reddish-brown coloring). They are organic-walled and may be extracted from a variety of sedimentary rock types. Dinoflagellates are significant because some species produce neurotoxins that can kill fish and accumulate in shellfish that are filter feeders. People who eat fish or shellfish that have ingested large quantities of dinoflagellates can develop a condition known as **paralytic shellfish poisoning** that can lead to long-lasting paralysis of the chest and abdominal muscles and can be fatal. Dinoflagellates are also known for their **bioluminescence** or flashes of light in ocean water or beach sand at night when they are disturbed.

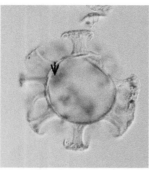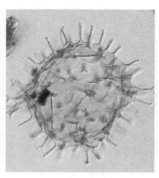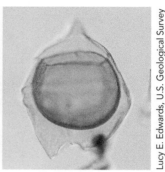

Lucy E. Edwards, U.S. Geological Survey

A. Dinoflagellates.

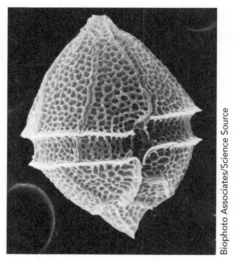

Biophoto Associates/Science Source

B. *Amphidoma nucula,* a dinoflagellate. Scanning electron micrograph, 30 micrometers long, 22 micrometers wide.

Figure 9.5 Dinoflagellates.

Coccolithophores (Calcareous Nannoplankton) (Figure 9.6)

You will not see actual specimens of coccolithophores in lab because they are too small to see without an electron microscope or an oil immersion lens on a compound microscope. The organism is called a **coccolithophore**, and it lives within a sphere of individual circular plates called **coccoliths**. This group belongs to the Haptophytes or golden-brown algae.

Geologic Range

Early Jurassic to Recent (Quaternary).

Shell Composition

Calcite.

Size

0.002 to 0.02 mm (2–20 μm). They are so small that they must be studied with an electron microscope or oil immersion lens on a compound microscope. You cannot see them with a regular light microscope.

Significance

Coccolithophores are the base of the marine food chain (phytoplankton) and are useful in biostratigraphy. Coccoliths are the major constituent of **chalk.**

Morphology

The organism is spherical to subspherical and covered by circular plates called **coccoliths.** Coccoliths can resemble a button or a daisy with petal-like ornamentation around the edge.

Environment

Marine only; exclusively planktonic.

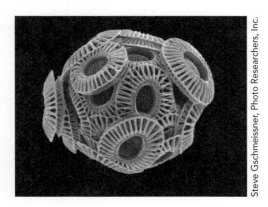

Steve Gschmeissner, Photo Researchers, Inc.

A. Scanning electron microscope image of a coccolithophore.

Pamela Gore

B. Coccolith in chalk, viewed with a scanning electron microscope.

C. Coccolithophore *(left)* and coccolith, a single plate *(right).*

Figure 9.6 Coccolithophores

Opisthokonta

The Opisthokonta include multicellular organisms such as animals (called **Metazoa** in the tree of life in Figure 9.1) and fungi, as well as various groups of protists. Small or microscopic parts of multicellular animals that you might see in lab include **sponge spicules, ostracodes,** and **conodonts**. Fungi and other Opisthokont protists will not be studied in this lab. All of the Opisthokonts we will see in lab are animals.

Sponge Spicules (Figure 9.7)

Sponges belong to the animal group **Porifera** (see Laboratory 10: Invertebrate Macro-fossils and Classification of Organisms). **Sponge spicules** are microscopic needle-like and multirayed skeletal elements secreted by sponge cells. Some groups of sponges secrete spicules of calcium carbonate and other groups secrete spicules of silica, or organic fibers.

Geologic Range
Cambrian to Recent (Quaternary).

Composition
Calcareous or siliceous.

Size
Varies.

Significance
Siliceous spicules can accumulate to form chert.

Morphology
Shapes of sponge spicules vary but may be needle-like (monaxon or one axis), three pointed (triaxon), four pointed (tetraxon), or shaped like a jack (from the game of ball and jacks) with six radiating needle-like points or rays (hexactine). They may also be curved.

Environment
Sponges live attached to the sea floor (sessile benthos). Most are marine.

A. Assorted types of sponge spicules.

Pamela Gore

B. *Astreospongia*, a siliceous sponge with six-rayed spicules (Silurian).

Figure 9.7   Sponge Spicules.

Ostracodes (Figure 9.8)

Ostracodes (pronounced "*ah*-struh-cods") are a type of shrimplike animal or crustacean belonging to the group **Arthropoda** (see Laboratory 10: Invertebrate Macrofossils and Classification of Organisms). Ostracodes have two valves or shells that are hinged at the top; they resemble a tiny kidney bean. The valves are made of calcium carbonate and are commonly covered with complex ornamentation consisting of bumps, pits, and ridges.

Geologic Range
Cambrian to Recent (Quaternary).

Shell composition
Calcareous (some organic).

Size
0.5 to 3.0 mm; some are larger.

Significance
Ostracodes are useful in biostratigraphy and paleoenvironmental interpretation.

Morphology
An ostracode is a microscopic shrimplike animal inside a clamlike shell consisting of two valves (shell halves) with a dorsal hinge.

Environment
Marine and nonmarine (fresh, brackish, and hypersaline); most are benthic (bottom dwellers).

A. Sketch of ostracodes.

B. Ostracodes in Late Triassic nonmarine shale, North Carolina. (Scale in millimeters.)

C. Light microscope composite all-in-focus image of ostracode genus *Bradleya*.

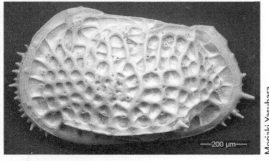

D. Scanning electron microscope photomicrograph of ostracode genus *Poseidonamicus*.

Figure 9.8 Ostracodes.

Conodonts (Figure 9.9)

Conodonts are small toothlike hard parts from an extinct animal in the group **Chordata**, to which vertebrates and some other organisms belong (see Laboratory 12: Evolution of the Vertebrates). Conodonts are made of apatite, a phosphate mineral that is similar in composition to our teeth and bones. They were part of a feeding apparatus, but not teeth, because they were covered with tissue. The name *conodont* means "cone-tooth," in reference to their shape. The individual hard parts are sometimes referred to as **conodont elements**, to distinguish them from the conodont animal.

The conodont animal was an elongated, soft-bodied, fishlike or wormlike chordate (resembling an eel) with a well-defined head, a notochord, and a distinct tail with fins. The conodont animal was not determined until the mid-1990s, although conodont elements have been known since the 1850s.

Geologic Range

Cambrian to Late Triassic. The conodont animals are extinct.

Composition

Phosphate (calcium fluorapatite).

Size

Most are 0.5 to 1.5 mm; some are up to 10 mm, and some are as small as 0.1 mm.

Significance

Conodonts are useful in biostratigraphy and marine paleoenvironmental interpretation. Their color is a good indicator of the temperature to which the enclosing rock has been subjected. (This is important in determining whether oil or gas may be present in the rock.) The darker the color of the conodont, the higher the temperature to which the rocks have been heated.

Morphology

Conodonts are parts of a larger organism, and resemble cone-shaped teeth or have bars with rows of toothlike denticles or irregular knobby plates called *platform elements*.

Environment

The conodont animal was marine and free-swimming.

A. Sketches of three types of conodonts. *From left to right*: a bar with a row of toothlike denticles, a simple cone-shaped element, and a platform element.

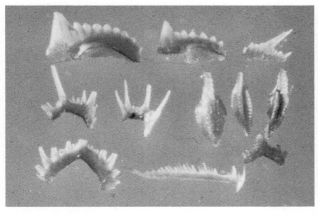 

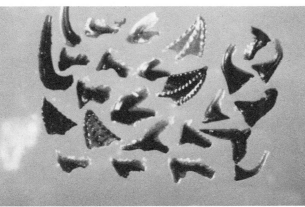 

Photos courtesy of Dr. Anita G. Harris, U.S. Geological Survey

B. Photomicrographs of conodonts. (1) Condonts from the Upper Pennsylvanian of Ohio.
(2) Condonts from the Lower Mississippian and redeposited Upper Devonian of Missouri.
(3) Conodonts from the Tuckaleechee Cove Window Limestone, lower Middle Ordovician, Tennessee.
(4) Platform conodonts from the lower Upper Devonian Genesee Formation, Yates County, New York.

Figure 9.9   Conodonts

Archaeplastida

The Archaeplastida have **plastids** (a type of cell organelle) with chlorophyll, and they are photosynthetic. Plastids probably originated through endosymbiosis with a cyanobacterium hundreds of millions of years ago. The Archaeplastida include the plants, green algae (chlorophytes such as the calcareous algae), charophytes, red algae (rhodophytes), and others.

Plant Pollen and Spores (Figure 9.10)

Plants have microscopic unicellular reproductive structures called **spores** and **pollen** that are sometimes preserved as fossils in sedimentary rocks, particularly fine-grained, gray to black sedimentary rocks. Spores and pollen are made of organic material, and they may be extracted from a variety of types of sedimentary rocks.

Geologic Ranges
Spores (from algae, fungi, mosses and ferns): Silurian to Recent (Quaternary).
Pollen from gymnosperms (conifers, ginkgoes): Pennsylvanian to Recent (Quaternary).
Pollen from angiosperms (flowering plants): Cretaceous to Recent (Quaternary).

Composition
Organic material (sporopollenin).

Size
0.02 to 0.08 μm; some are up to 0.2 mm.

Significance
Pollen and spores are useful in biostratigraphy and in paleoenvironmental and paleoclimatic interpretations.

Morphology
Globular or spheroidal. Some pollen is shaped like Mickey Mouse ears.

Environment
Pollen and spores come from land plants. Pollen and spores are most commonly found in freshwater deposits (continental sedimentary environments), but they can also be found in transitional sedimentary environments.

A. Pollen from a variety of common plants. The bean-shaped grain at lower left is about 50 μm in diameter.

Courtesy of Dartmouth Electron Microscope Facility

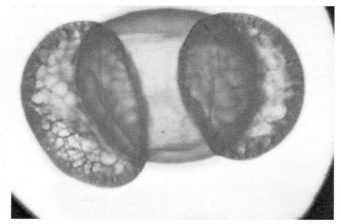

B. Pollen grains from a fir tree. This grain is about 125 microns wide, and has been stained red for microscopical viewing. (From Levin, H., 2013, *The Earth Through Time* (10th edition), figure 6.31(a), p. 149. This material is reproduced with permission of John Wiley & Sons, Inc.)

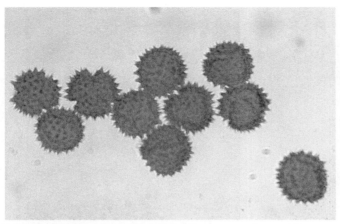

C. Ragweed pollen. These pollens are about 25 microns in diameter and have been stained red for microscopical viewing. (From Levin, H., 2013, *The Earth Through Time* (10th edition), figure 6.31b, p. 149. This material is reproduced with permission of John Wiley & Sons, Inc.)

Figure 9.10 Plant Pollen and Spores.

D. Yellow pollen from wind-pollinated trees, such as pines, coats all surfaces in the spring in Atlanta, Georgia (such as this car), and would be an obvious contributor to sediments, particularly in lakes with quiet anoxic water. Pollen and spores are commonly found in gray and black organic-rich shales.

Calcareous algae (Figure 9.11)

The calcareous algae are **chlorophytes**, a type of green algae. Calcareous algae include the modern *Halimeda, Penicillus,* and *Acetabularia* and the extinct *Receptaculites.* The calcareous algae are macroscopic (several centimeters or larger), erect bottom dwellers in warm, clear, shallow tropical marine waters. They are not microfossils, but they are single-celled organisms. Calcareous algae are significant contributors to lime muds and other carbonate sediment.

Geologic Range
As a group, they are Early Ordovician to Recent (Quaternary). The geological range of *Halimeda* is Cretaceous to Recent (Quaternary).

Composition
Aragonite (calcium carbonate).

Size
Calcareous algae can grow to 10 to 15 cm or more. A few are meters across.

Significance
Calcareous algae are major contributors to marine sediments; many modern forms disaggregate into aragonite needle muds that contribute to the formation of micrite or fine-grained limestone.

Morphology
Halimeda is composed of segments somewhat similar in appearance to a Christmas cactus. Each segment resembles a three-fingered glove; new segments arise from the tip of each finger, leading to a branching shape. The whole organism (and sometimes large clusters of calcareous algae on the sea floor) consists of a single cell with multiple nuclei, attached by filament-like threads passing through the mud. *Penicillus* resembles a shaving brush. *Acetabularia*, with a radially symmetrical disk at the top of a stem, is also single-celled. The extinct *Receptaculites* (Early Ordovician to Permian) resembles a sunflower with a spiral pattern.

Environment
Marine, tropical to subtropical.

A. *Halimeda,* a type of calcareous algae from Florida Bay. Specimen is about 5 cm high.

B. *Penicillus*, a type of calcareous algae from Florida Bay. The large structure at the bottom is rootlike and in life is below the sediment surface. (Scale in centimeters.)

C. *Acetabularia,* a type of calcareous algae. The large structure at the bottom is rootlike and in life is below the sediment surface. (Scale in centimeters.)

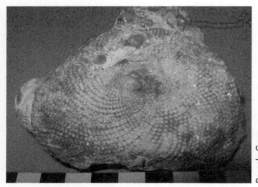

D. *Receptaculites,* extinct calcareous algae (Early Ordovician to Permian). (Scale in centimeters.)

Figure 9.11 Calcareous algae

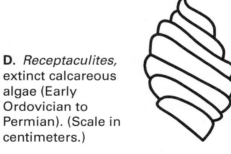

Figure 9.12 Sketch of a charophyte oogonium, about 0.5 mm in diameter.

Charophytes

Charophytes are another type of green algae, also known as stoneworts. Their female reproductive structures (oogonia) are encased in calcium carbonate that can be preserved as microfossils. The fossilized oogonia (called gyrogonites) are small ovoid bodies encircled by spiral ridges and grooves (Figure 9.12).

Geologic range
Upper Silurian to Recent (Quaternary).

Composition
Calcium carbonate.

Size
The oogonia are about 0.5 mm in diameter. The entire stonewort may be 30 cm or more in length.

Significance
Charophytes are the closest living relatives to the ancestors of land plants. They can be a contributor to lake sediment. They are useful for correlation in freshwater deposits.

Morphology
Charophytes have an ellipsoidal shape with spiral ridges or grooves.

Environment
Charophytes mostly inhabit quiet fresh to brackish (low salinity) water: lakes, wetlands, rivers, streams, estuaries.

Coralline Algae

The coralline algae are a type of rhodophyte (or red algae) (Figure 9.13). They are important tropical reef formers because they secrete calcium carbonate hard parts, as well as encrusting and binding other reef organisms. Their pink to red calcium carbonate grains are a significant contributor to tropical carbonate sands (Figure 9.14).

Geologic Range

Red algae range from Proterozoic to Recent (Quaternary). The true coralline algae range from Jurassic to Recent (Quaternary).

Composition

Calcite, some with high-magnesium calcite.

Size

Several centimeters or more.

Significance

The oldest identified multicellular eukaryote on record is a type of red algae. They are important in reef construction. Red algae can be used as an indicator of water depth and temperature.

Morphology

The red algae include both thin, delicate branching forms and red encrusting forms. Most are multicellular.

Environment

Mostly marine in warm, tropical waters.

Figure 9.13   Branching and red encrusting forms of red coralline algae. (Scale in centimeters.)

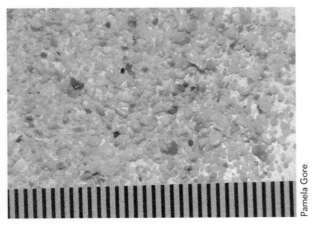

Figure 9.14   Carbonate sand. The pink to red grains are the remains of red algae. (Scale in millimeters.)

SELECTED REFERENCES

Adl SM, Simpson AG, Farmer MA, et al. (2005). The new higher level classification of eukaryotes with emphasis on the taxonomy of protists. Journal of Eukaryotic Microbiology 52(5): 399–451. http://onlinelibrary.wiley.com/doi/10.1111/j.1550-7408.2005.00053.x/abstract.

Burki F, Shalchian-Tabrizi K, Minge M, et al. (2007). Phylogenomics reshuffles the eukaryotic supergroups. PloS ONE 2(8):e790. http://www.plosone.org/article/info:doi%2F10.1371%2Fjournal.pone.0000790

Keeling P, Leander BS, Simpson A. Eukaryotes. Tree of Life Web Project. http://tolweb.org/Eukaryotes.

National Center for Biotechnology Information, U.S. National Library of Medicine. Taxonomy. http://www.ncbi.nlm.nih.gov/guide/taxonomy/

Olander N. Tree of Life. http://tellapallet.com/tree_of_life.htm

Science. A Tree of Life (2003). http://www.sciencemag.org/feature/data/tol/.

Tree of Life (Special Issue). Science 300(5626):1605–1832. http://www.sciencemag.org/content/vol300/issue5626/index.dtl.

Microfossils and Introduction to the Tree of Life Exercises

PRE-LAB EXERCISES

1. List the composition of the hard parts for each group of microfossils.

Microfossil Group	Composition of Hard Parts
a. Foraminifera	
b. Radiolaria	
c. Diatom	
d. Coccolithophores	
e. Dinoflagellates	
f. Ostracodes	
g. Conodonts	
h. Spores	
i. Pollen	
j. Sponge spicules	

2. What type of rock will these microfossils accumulate to form?

Microfossil Group	Rock formed by these Microfossils
a. Diatoms	
b. Coccolithophores	

3. Into which eukaryote supergroup would you place the following organisms?
A = Amoebozoa, B = Opisthokonta, C = Rhizaria, D = Archaeplastida,
E = Excavata. Put the letter in the blank.

Microfossil Group	Eukaryote Supergroup
a. Foraminifera	
b. Diatom	
c. Coccolithophore	
d. Radiolarian	
e. *Halimeda*	
f. Sponge	
g. Dinoflagellate	

Microfossil Group	Eukaryote Supergroup
h. Pine tree	
i. Ostracode	
j. Conodont	
k. Charophyte	
l. Coralline algae	
n. Mushroom	
o. Human	

4. Microfossils are useful for biostratigraphy or correlating sedimentary rocks and determining their ages. In the table below, indicate the geologic ranges of each microfossil group by placing an x in the cells corresponding to the geologic periods in which they are found.

Microfossil Group	Paleozoic							Mesozoic			Cenozoic		
	Ꞓ	O	S	D	M	ℙ	P	Ƭ	J	K	ℙ	N	Q
Planktonic foraminifera													
Benthic foraminifera													
Radiolarians													
Diatoms													
Coccolithophores													
Dinoflagellates													
Ostracodes													
Conodonts													
Sponge spicules													
Spores													
Pollen (gymnosperm)													
Pollen (angiosperm)													
Calcareous algae													
Charophytes													

Key: Ꞓ = Cambrian, O = Ordovician, S= Silurian, D = Devonian, M = Mississippian, ℙ = Pennsylvanian, P = Permian, Ƭ = Triassic, J = Jurassic, K = Cretaceous, ℙ = Paleogene, N = Neogene, Q = Quaternary.

5. As you can see from your completed table in exercise 4, some groups of organisms have long geologic ranges, and others have much shorter geologic ranges.

a. Which groups in the table above have the longest geologic ranges?

_____

b. Which groups have short geologic ranges? _____

_____

c. Which group is extinct? _____

6. Individual fossil species have much shorter geologic ranges than the group as a whole. If you were to find an ostracode preserved in a sedimentary rock, it might be of any age from Cambrian to Recent. However, if you could identify the particular species of ostracode, it might give you the age of the rock within a few million years. Commonly, assemblages of several types of fossils are found together in a rock. This can also help you determine the age of the rock more closely. For example, if you had an ostracode fossil and a planktonic foram fossil together in a rock, what would this tell you about the age of the rock? Because planktonic forams first appear in the fossil record in the Jurassic, this means that the rock could be no older than the Jurassic; however, it could be of any age between Jurassic and Quaternary. If you can identify the species of fossil, that would help you limit the age of the rock to a narrower time interval.

a. What age would a rock be if it contains fossils of diatoms and coccolithophores? (Give period names.)

_____

b. What age would a rock be if it contains fossils of dinoflagellates and conodonts? (Give period names.)

_____

c. What age would a rock be if it contains fossils of both gymnosperm and angiosperm pollen? (Give period names.)

_____

d. What age would a rock be if it contains sponge spicules, radiolarians, and diatoms? (Give period names.)

_____

e. Would you expect to find conodonts and planktonic foraminifera together in a rock? Explain why or why not.

_____

7. Microfossils are useful for paleoenvironmental interpretation. Put an x in the cell for the environments in which the microfossil groups lived.

Microfossil Group	Continental (nonmarine/freshwater)	Marine
Foraminifera		
Radiolaria		
Diatom		
Coccolithophores		
Dinoflagellates		
Ostracodes		
Conodonts		
Spores		
Pollen (gymnosperm)		
Pollen (angiosperm)		
Calcareous algae		
Charophytes		
Coralline algae		

8. A particular sequence of sedimentary rock layers is found to contain the microfossils listed in the table below.

    a. List the depositional environment (marine, continental (nonmarine/freshwater), or mixed) of the groups in each layer, in the table.

    b. What age would each layer of rock be that contains this assemblage of microfossils? Give period name(s) for each layer. Make sure that *all* of the microfossils lived during the period names you list.

Position in Vertical Sequence	Microfossil Groups	Depositional Environment (Marine, Continental (Nonmarine/Freshwater), Mixed)	Geologic Periods Where All Might Be Found Together
Layer A (Top)	Angiosperm pollen		
Layer B	Angiosperm pollen and ostracodes		
Layer C	Ostracodes, angiosperm pollen, and diatoms		
Layer D	Planktonic foraminifera and ostracodes		
Layer E (Bottom)	Coccoliths, foraminifera, and radiolaria		

    c. Which layers (give letters) contain microfossil groups characteristic of *marine only* environments?

_____

    d. Which layers (give letters) contain microfossil groups characteristic of *continental (nonmarine/freshwater) only* environments?

_____

    c. From what you know about the depositional environments that these groups inhabit, is this a *transgressive sequence* or a *regressive sequence*?

_____

    d. What age (give period names) would the rock unit be that contains these *five layers* of sedimentary rock, and this *entire* assemblage of microfossils? The answer will be more than one geological period name. Review the geological time chart if you do not remember the period names.

_____

9. Which microfossil groups can be found living in both marine and continental (nonmarine/freshwater) environments? (List three.)

_____

_____

_____

10. Thinking question: Explain how it might be possible for pollen and spores to become preserved in marine sediments.

11. In some areas of the sea floor, foraminifera tests (shells) are accumulating at a rate of 1 cm per 1,000 years. How thick a deposit could have accumulated during the 65 million years (65,000,000 years) of the Cenozoic Era? (Use your calculator.)

_____ cm

LAB EXERCISES

1. Identify the microfossils in the chart below.

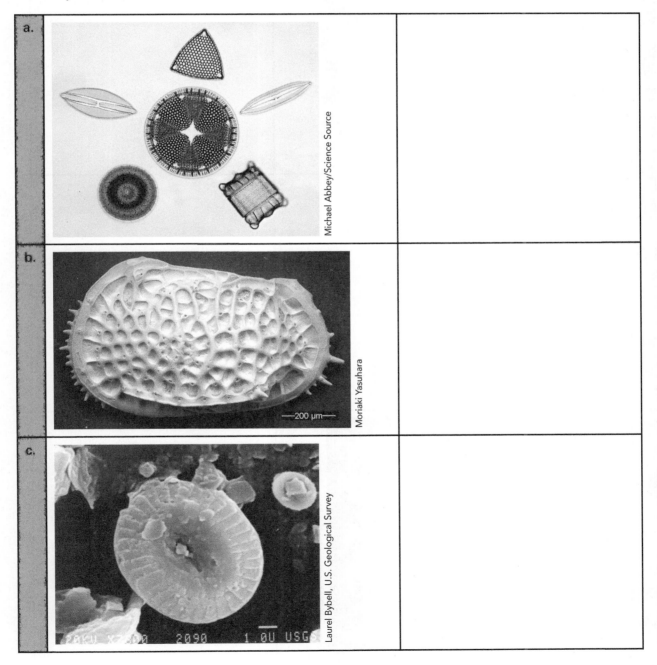

d.

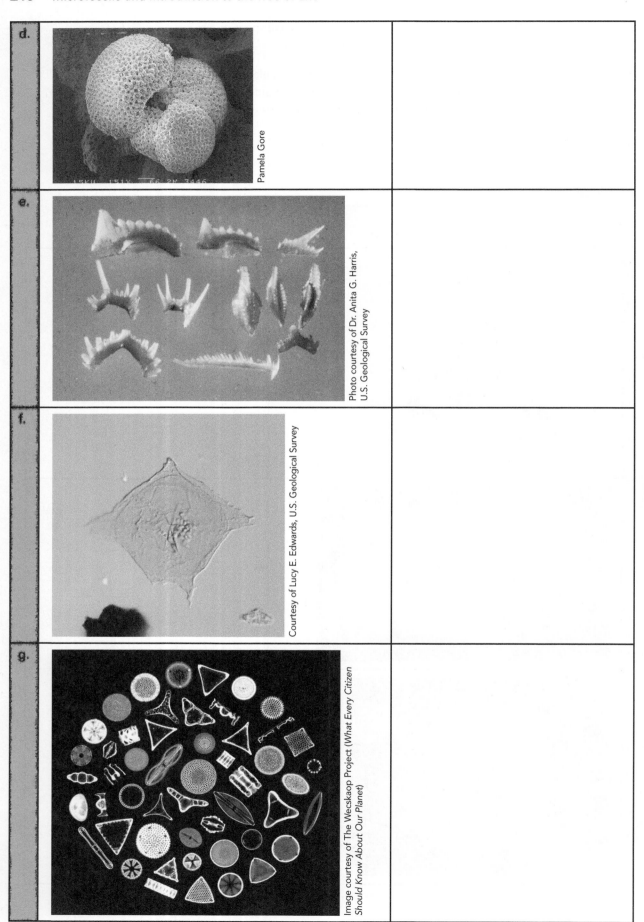

Pamela Gore

e.

Photo courtesy of Dr. Anita G. Harris,
U.S. Geological Survey

f.

Courtesy of Lucy E. Edwards, U.S. Geological Survey

g.

Image courtesy of The Wecskaop Project (*What Every Citizen
Should Know About Our Planet*)

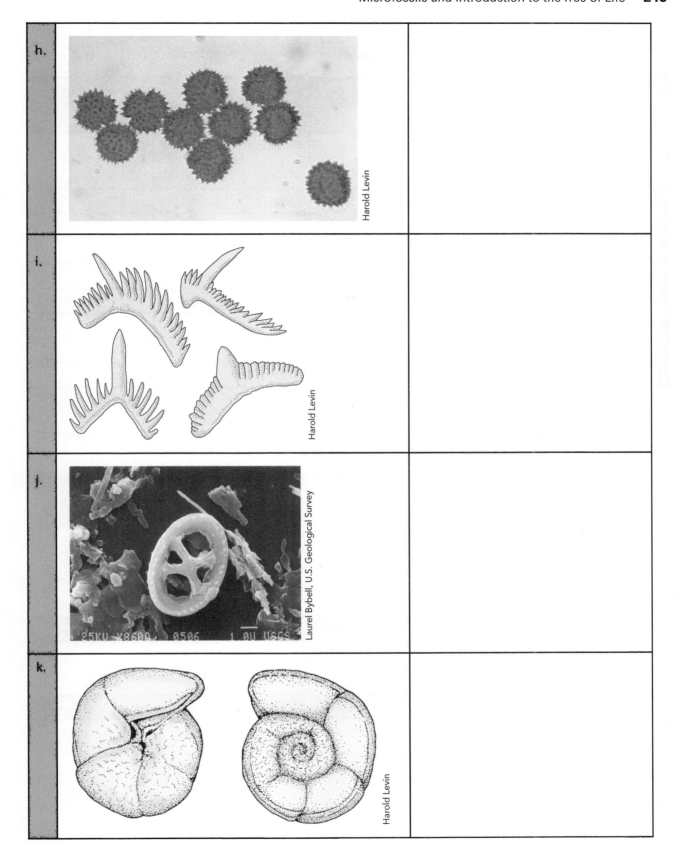

h.

Harold Levin

i.

Harold Levin

j.

Laurel Bybell, U.S. Geological Survey

'25KU ×8600. 0506   1.0U USGS

k.

Harold Levin

2. Examine the microfossil slides or vials provided by your instructor using a microscope or hand lens and fill in the chart below.

	Type of Microfossil	Supergroup and any other Taxonomic Rank available	Composition: Calcite/Aragonite, Silica, Phosphate, Organic	Unicellular or Multicellular
Example	Conodont	Opisthokont Animal Chordate	Phosphate	Multicellular
a.				
b.				
c.				
d.				
e.				
f.				
g.				
h.				

3. Examine specimens of large foraminifera (nummulitids and fusulinids) provided by your instructor. Draw a sketch of each in the chart below.

Nummulitid foraminifera	Fusulinid foraminifera

4. Examine and sketch additional specimens provided by your instructor.

Sketch	Identity

Sketch	Identity

Balanced Rock, Garden of the Gods, Colorado Springs, Colorado.

Invertebrate Macrofossils and Classification of Organisms

In this lab, you will examine fossil animals known as invertebrates. **Invertebrates** are animals without backbones. **Fossils** are the prehistoric remains of ancient organisms. **Macrofossils** are fossils large enough to be studied without the aid of a microscope. Fossils are useful for biostratigraphy or correlating sedimentary rocks and determining their ages, as well as for interpreting depositional environments.

CLASSIFICATION OF THE ANIMALS

Taxonomy is the system of naming and classifying (or grouping) organisms. Animals are traditionally classified using an artificial system of grouping by similar anatomical features or body construction. The major groups of animals are called phyla (singular, phylum). Phyla are subdivided into smaller and smaller groups on the basis of how closely the animals resemble one another:

Kingdom Animalia

Phylum

Class

Order

Family

Genus

Species

This is known as rank-based scientific classification or Linnaean taxonomy (named for Carolus [or Carl] Linnaeus, 1707–1778). The genus and species names are typically italicized. The name of the genus is capitalized, but the name of the species is not (e.g., *Homo sapiens*). This system of naming organisms is called **binomial nomenclature** (meaning a two-part name, referring to genus and species).

In many textbooks, organisms are grouped into three Domains and six Kingdoms:

> **Box 10.1   Classification of organisms**
>
> 1. **Domain Bacteria**
>   Kingdom Eubacteria (bacteria and cyanobacteria or blue-green algae)
> 2. **Domain Archaea**
>   Kingdom Archaebacteria (hyperthermophiles, halophiles, methanogens)
> 3. **Domain Eukaryota or Eukarya (eukaryotes)**
>   Kingdom Animalia (animals)
>   Kingdom Plantae (plants)
>   Kingdom Protista (foraminifera, radiolarians, diatoms, dinoflagellates)
>   Kingdom Fungi (mold, mushrooms, yeast, fungus)

Organisms from each of the three domains are known as fossils. The Bacteria and Archaea are prokaryotes. **Prokaryotes** are single-celled organisms with cells that do not have a nucleus (called **prokaryotic cells**). The **Eukaryota** (animals, plants, protists, and fungi) have cells with nuclei (called **eukaryotic cells**). The prokaryotic cells are considered more primitive than eukaryotic cells. Oddly enough, studies of genes and biochemistry suggest that Archaea are more closely related to Eukaryota than to Bacteria, although other aspects of their biochemistry are unique. Many Archaea inhabit extreme environments such as hot springs, geysers, and black smokers on the sea floor, but they also inhabit a wide variety of other environments.

Eukaryotes (multicellular organisms, and single-celled organisms with a nucleus) are divided into six supergroups based on genetic similarities and differences. This terminology is new (since about 2003), and most textbooks, biologists, and college faculty are still catching up to the rapid changes in classification. The six eukaryote supergroups are:

1. **Rhizaria:** foraminifera, radiolaria, and others
2. **Chromalveolata:** diatoms, dinoflagellates, coccolithophores, and others
3. **Opisthokonta:** animals, fungi, and others
4. **Archaeplastida:** plants, green algae, calcareous algae, red algae, and others
5. **Amoebozoa:** amoebas and slime molds
6. **Excavata:** some flagellate species (*Euglena, Giardia, Trichomonas, Leishmania*)

The animals (both invertebrates and vertebrates) belong to the supergroup **Opisthokonta**. The word *opisthokonta* means "rear pole," referring to a flagellum attached to the posterior of the cell (e.g., a sperm cell). Some studies have suggested genetic similarities between some of the supergroups; the Opisthokonta and the Amoebozoa are sometimes combined into the **Unikonts**.

Animals and plants have some characteristics in common, but they also have basic differences. Most **plants** manufacture their own food by photosynthesis (i.e., they are producers or **autotrophs**). There are a few exceptions, such as some unusual plants that are parasitic and not photosynthetic (e.g., dodder and Indian pipe). Furthermore, plants have no freedom of movement. **Animals** cannot manufacture their own food; they must eat (i.e., they are consumers, or **heterotrophs**). Animals have at least some freedom of movement; however, many types of animals live attached to rocks or other organisms (e.g., barnacles, oysters, corals).

MOLECULAR PHYLOGENETICS: THE LATEST CLASSIFICATION SYSTEM

The latest classification of organisms is based on gene sequencing and molecular data from DNA. High-powered computers are used to compare selected genes or the entire genetic sequence of many different types of animals and protists. This is called **molecular phylogenetics**. The new classification scheme recognizes groups of animals that are related by evolution, as shown by genetic similarities, and confirms relationships that had been established earlier, based on anatomy. These molecular genetic studies are also supported by biochemical data and morphological data from ultrastructural research on cells using electron microscopes. Our understanding of the tree of life is currently in a state of flux, and these groupings will continue to be refined with further studies of molecular phylogenetics.

The terminology used for taxonomy or classification of organisms changed in 2005, and some traditional groupings of organisms have been abandoned by some scientists. The three domains remain, but instead of Kingdom, Phylum, Class, Order, and so on, there are now nameless ranks, with the ranked hierarchy indicated by indentation and dots, as in the taxonomic hierarchy in Box 10.2. This approach has the advantage of flexibility and ease of modification with rapidly occurring advances in the study of gene sequencing. It also avoids having to deal with intermediate taxonomic units such as subphylum, superclass, and subclass. This is all new since about 2005, and textbooks, biologists, geologists, and college faculty are still catching up to the rapid changes in classification. There will undoubtedly be some resistance to letting go of older, more comfortable, rank-based Linnaean taxonomic terminology. Expect changes to continue. Such is the progress of science.

The taxonomic hierarchy (Box 10.2) summarizes the invertebrate groups that are studied in this lab. (The vertebrates are included with the Chordates at the end of the list.) This list is not complete. A number of types of soft-bodied animals leave no significant fossil record, and they have been omitted here, such as the many types of worms. All of these groups are further subdivided, but these are they only groups that you need to know for this class.

The animals that we will study are divided into three groups: **Porifera** (sponges), **Radiata** or radially symmetrical, roughly circular animals (such as jellyfish), and the **Bilateria** or bilaterally symmetrical animals with a front and back end (anterior and posterior) and a top and bottom side (dorsal and ventral). In general, the body plan is basically a tube. Bilateria include animals such as insects and mammals. Porifera diverged from the other animal groups early in evolutionary history; they have several types of cells, but the main difference is that their cells are not organized into tissues.

Animals can be grouped into **clades**, which are groupings of organisms based on genetic similarities, reflecting evolutionary relationships. The **Bilateria** or bilaterally symmetrical animals are divided into **protostomes** (meaning "mouth first") and **deuterostomes** (meaning "mouth second"), which differ in their embryonic development. In general, this relates to whether the mouth or the anus formed first in the embryo. Protostomes include bryozoans, brachiopods, molluscs and arthropods. Deuterostomes include echinoderms and chordates.

The two major clades of bilaterally symmetrical protostomes are the Ecdysozoa and the Lophotrochozoa. **Ecdysozoa** are animals that shed their "skin" or exoskeleton (arthropods such as insects, shrimp, and crabs). **Lophotrochozoa** include animals with a lophophore (a ring of specialized tentacles around the mouth) such as bryozoans and brachiopods, and another group that includes molluscs and worms.

The dots and indentation in the outline in Box 10.2 indicate the nameless rank in the taxonomic hierarchy. The letters and numbers were added to correspond to those used in the outline in this lab. This information could also be presented in the form of a branching tree diagram or tree of life, which you can see online.[1]

In the taxonomic hierarchy (Box 10.2), the units that were phyla are shown in bold letters with Roman numerals. You can see that what we called "phyla" are actually at different levels in the taxonomic hierarchy (as indicated by dots and indentation.) *You need to know these terms, and the names of most of the subgroups under them. These are the types of fossils that you will be studying in the lab.*

---

[1] Olander N. Tree of Life. http://tellapallet.com/tree_of_life.htm

Box 10.2 Taxonomic Hierarchy

Domain Eukaryota

Supergroup Opisthokonta

- Animalia or Metazoa (multicellular organisms)
- • **I. Porifera** (sponges)
- • *Radiata* (animals with radial symmetry)
  - • • **II. Cnidaria** (corals, sea anemones and jellyfish)
    - • • • A. Anthozoa (sea anemones and corals)
      - • • • • 1. Rugose corals
      - • • • • 2. Tabulate corals
      - • • • • 3. Scleractinian corals
    - • • • B. Scyphozoa (jellyfish)
- • *Bilateria* (bilaterally symmetrical animals)
  - • • • *Protostomia* (A term relating to mouth formation in embryonic development. In the embryo, the first opening becomes the mouth and the second opening becomes the anus.)
    - • • • *Lophotrochozoa*
      - • • • • **III. Bryozoa**
      - • • • • **IV. Brachiopoda**
        - • • • • • A. Inarticulata
        - • • • • • B. Articulata
      - • • • • **V. Mollusca**
        - • • • • • A. Bivalvia (clams, oysters, scallops, mussels)
        - • • • • • B. Gastropoda (snails and slugs)
        - • • • • • C. Cephalopoda (squid, octopus, cuttlefish)
          - • • • • • • 1. Nautiloidea (*Nautilus)*
          - • • • • • • 2. Ammonoidea (ammonoids)
          - • • • • • • 3. Coleoidea
            - • • • • • • • a. Belemnoidea (belemnites)
            - • • • • • • • b. Neocoleoidea
              - • • • • • • • • 1. Sepiida (cuttlefish)
              - • • • • • • • • 2. Teuthida (squids)
              - • • • • • • • • 3. Octopoda (octopus)
        - • • • • • D. Scaphopoda (tusk shells)
        - • • • • • E. Monoplacophora
        - • • • • • F. Polyplacophora (chitons)
    - • • • *Ecdysozoa* (moulting or shedding animals)
      - • • • • **VI. Arthropoda**
        - • • • • • A. Trilobita (triobites)
        - • • • • • B. Chelicerata
          - • • • • • • 1. Merostomata
            - • • • • • • • a. Xiphosura (horseshoe crabs)
            - • • • • • • • b. Eurypterida (eurypterids)
          - • • • • • • 2. Arachnida (spiders)
        - • • • • • C. Crustacea (crustaceans)
          - • • • • • • 1. Ostracoda (ostracodes)
          - • • • • • • 2. Cirripedia (barnacles)
          - • • • • • • 3. Malacostraca

•••••••• a. Decapoda (shrimp, crabs)

•••••••• b. Isopoda (sowbugs, pillbugs or roly-polys)

••••••• 4. Branchiopods (brine shrimp)

•••••••• a. Conchostraca (clam shrimp)

•••••••• b. Notostraca (tadpole shrimp, *Triops*)

•••••••• c. Cladocera (water fleas)

•••••••• d. Anostraca (brine shrimp)

•••••• D. Hexapoda

••••••• 1. Insecta

••• *Deuterostomia* (In the embryo, the first opening becomes the anus and the second opening becomes the mouth.)

•••• **VII. Echinodermata**

••••• A. Crinoidea (crinoids)

••••• B. Blastoidea (blastoids)

••••• C. Asteroidea (starfish)

••••• D. Ophiuroidea (brittle stars)

••••• E. Echinoidea (sand dollars and sea urchins)

••••• F. Holothuroidea (sea cucumbers)

•••• **VIII. Hemichordata** (graptolites)

••••• A. Graptolithina (graptolites)

•••• **IX. Chordata** (vertebrates; see Laboratory 12: Evolution of the Vertebrates)

DOMAIN EUKARYOTA: SUPERGROUP OPISTHOKONTA: ANIMALIA (METAZOA OR MULTICELLULAR ORGANISMS)

I. Porifera: Sponges (Phylum Porifera) (Figure 10.1)

Name

Porifera means "pore-bearing." Exterior is covered by tiny pores.

Chief Characteristics

Globular, cylindrical, conical or irregular shape. Interior may be hollow or filled with branching canals. Solitary or colonial. Skeletal elements are called **spicules**, and they may be separate or joined.

Living forms contain a variety of cell types (amoeboid cells, choanocytes or flagellated collar cells, and others) that function more or less independently and can transform into other types as needed. The cells are not organized into tissues.

Geologic Range

Cambrian to Recent (Quaternary).

Types of Sponges and their Composition

Silicispongia: Usually with silica spicules

Hexactinellida: Silica spicules with six points (Class Hexactinellida)

Demospongiae: Skeletons composed of an organic material called **spongin**, sometimes with silica spicules (Class Demospongiae). Dominant group of living sponges.

Calcispongia: Calcium carbonate (calcite or aragonite) spicules (Class Calcarea)

Mode of Life

Attached to the sea floor (sessile benthos). Most are marine.

A. Living sponge.

B. Dried modern sponge with skeleton composed of an organic material called spongin. (Pencil with centimeter scale.)

C. *Astraeospongia,* fossil sponge (Silurian). Note the six-rayed star-shaped spicules. (Scale in centimeters.)

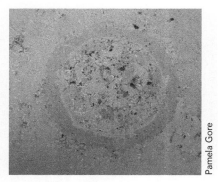

D. Transverse section of a fossil sponge. Solnhofen Limestone (Jurassic). (About 10 cm in diameter.)

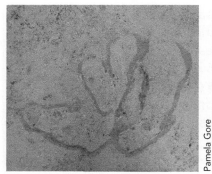

E. Longitudinal section through a cluster of fossil sponges; note openings at top. Solnhofen Limestone (Jurassic). (About 12 cm long.)

F. Longitudinal section through a fossil sponge; note opening at top. Solnhofen Limestone (Jurassic). (Pencil with centimeters for scale.)

Figure 10.1 Porifera. Modem and ancient sponges.

II. Cnidaria: Corals, Jellyfish, Sea Anemones (Phylum Cnidaria) (Figures 10.2, 10.3, 10.4, 10.5 and 10.6)

Name

Pronounced "nid-AREA," (the *c* is silent). Cnidaria are named for stinging cells called **cnidoblasts** (from *cnida* or *knide,* meaning "nettle" or "needle," and *blast,* meaning "cell").

Chief Characteristics

Radial symmetry. There are two basic body forms among the Cnidaria: a free-swimming jellyfish form, called a **medusa**, and a form that is attached to the sea floor, called a **polyp** (e.g., sea anemone or coral). Both are soft bodied, with a mouth surrounded by tentacles with stinging cells. The jellyfish belong to the **Scyphozoa** and the corals and sea anemones belong to the **Anthozoa**. Some corals have symbiotic algae called **zooxanthellae** living in their tissues. These algae are **dinoflagellates** (see Laboratory 9: Microfossils and Introduction to the Tree of Life).

Coral polyps may be **solitary** (one individual) or **colonial** (multiple individuals connected together). Coral polyps secrete a hard calcium carbonate skeleton called a **corallite**; each corallite is small (several millimeters to several centimeters or more in diameter). The corallites of **Rugose** and **Scleractinian** corals are roughly circular to hexagonal openings, with internal radial partitions called **septae** (singular: septum). **Tabulate** corals lack septae.

Cnidarians that form hard skeletal structures (such as corals) are most readily preserved as fossils; however, soft-bodied forms such as jellyfish are occasionally preserved. The only cnidarians that we will study in this lab are the corals.

Mode of Life

Corals and sea anemones live attached to the sea floor, a lifestyle called *sessile benthos*, meaning "seated bottom dweller". They live primarily in warm, shallow marine environments. Jellyfish are free-swimming marine organisms.

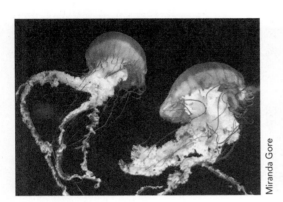

A. Jellyfish.

B. Jellyfish.

C. Jellyfish.

D. Fossil jellyfish, *Heimalora stellaris* (Upper Cambrian), Mt. Simon Complex, Wisconsin. Jellyfish is about 10 cm wide.

Figure 10.2 Modern and fossil Scyphozoa (jellyfish).

A. Coral reef ecosystem at Kure Atoll in the Northwestern Hawaiian Islands Marine National Monument.

B. Coral polyps.

C. Sea anemones.

D. Sea anemone.

Figure 10.3   Modern Anthozoa (corals and sea anemones).

A. Anthozoa: Corals and Sea Anemones (Class Anthozoa)

1. Rugose Corals (Order Rugosa) (Figure 10.4)

Chief characteristics

Most **rugose corals** are solitary and conical (shaped like ice-cream cones). **Septae** (radiating vertical partitions) are visible in the corallite (circular opening at the top of the cone). Some rugose corals are colonial and have roughly hexagonal corallites (such as *Hexagonaria*).

Composition

Calcite.

Geologic Range

Ordovician to Permian. All are extinct.

A. Solitary rugose corals. (Scale in centimeters.)

B. Side view of a colonial rugose coral. (Scale in centimeters.)

C. Colonial rugose coral, *Hexagonaria*. The corallites are hexagonal in shape, and the septae are clearly visible, radiating outward within each corallite. (Scale in centimeters.)

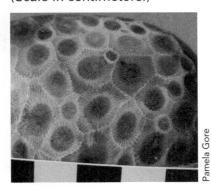

D. Colonial rugose coral, *Hexagonaria*, that has been rounded is commonly known as a Petoskey stone (Devonian), Michigan. This specimen has been polished, revealing the details of the septae. (Scale in centimeters.)

Figure 10.4 Rugose corals.

---

2. Tabulate Corals (Order Tabulata) (Figure 10.5)

Chief Characteristics

Tabulate corals are colonial and resemble honeycombs or wasps' nests. The corallites lack septae. In side view, tiny platforms called **tabulae** are visible within the corallites. The tabulae resemble a series of floors in a high-rise building.

Composition

Calcite.

Geologic Range

Ordovician to Permian. All are extinct.

A. Top view of a tabulate coral colony, showing the small corallites, which lack septae. (Scale in centimeters.)

B. Side view of a tabulate coral colony showing the tabulae. (Scale in centimeters.)

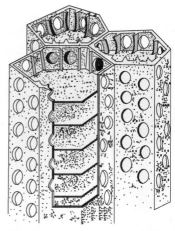

C. Skeletal structure of a tabulate coral, showing the tabulae, which resemble floors in a building. (From Levin, H., 2013, *The Earth Through Time* (10th edition), figure 12-25, p. 348. This material is reproduced with permission of John Wiley & Sons, Inc.)

D. Tabulate coral colony. (Scale in centimeters.)

Figure 10.5 Tabulate corals.

3. Scleractinian Corals (Order Scleractinia) (Figure 10.6)

Chief characteristics
Scleractinian corals are the modern corals, many of which are reef builders. Most are colonial, but some are solitary. Corallites have radiating septae. In some scleractinian corals, skeletal material is deposited between corallites (such as in *Astrhelia palmata*).

Composition
Aragonite.

Geologic Range
Triassic to Recent (Quaternary).

A. Detail of modern scleractinian coral showing corallites with septae. (Scale in centimeters.)

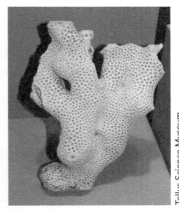

B. Fossil scleractinian coral, *Astrhelia palmata.*

C. Fossil scleractinian coral (Pleistocene), south Florida.

D. Detail of coral in **C,** showing septae.

Figure 10.6   Scleractinian corals.

III. Bryozoa: (Phylum Bryozoa) (Figure 10.7)

Name
Bryo means "moss", and *zoa* means "animal."

Chief Characteristics
Bryozoans are colonial. Many microscopic individuals live physically united adjacent to one another. Individuals are very tiny (a millimeter or less in diameter), just large enough to be seen with the unaided eye. Bryozoans may be distinguished from corals because the apertures in the skeleton are much smaller. Bryozoan colonies can resemble lace or a tiny net, may be delicately branching, finger-like, circular or dome shaped. *Archimedes* from the Mississippian has a corkscrew-like central axis with a fragile netlike colony around the outer edge that is typically broken off.

Composition
Calcite or aragonite.

Geologic Range
Ordovician to Recent (Quaternary).

Mode of Life

Widespread in marine environments. A few live in freshwater lakes and streams. Colonies may be encrusting, erect, or massive. Encrusting colonies have a thin layer covering a rock or shell. Erect colonies are attached at the bottom and stand up like a tree; some are branching, others are sheetlike. Massive colonies are compact and nodular.

Courtesy of Claire Fackler, National Oceanic and Atmospheric Administration National Marine Sanctuaries media library

A. A colorful living lacy bryozoan sits among other invertebrates and algae in the Channel Islands National Marine Sanctuary.

Pamela Gore

B. Modern bryozoan colony commonly known as hornwrack *(Flustra foliacea),* Nova Scotia. (Scale in centimeters.)

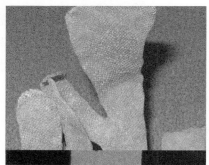

Pamela Gore

C. Detail of modern hornwrack bryozoan colony showing individual "boxes" in which each individual lived. (Scale in centimeters.)

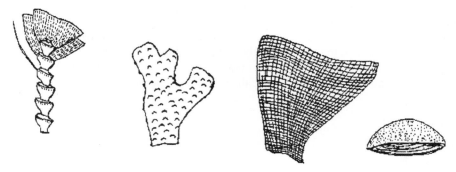

D. Sketches of a variety of types of bryozoans.

Figure 10.7 Bryozoans.

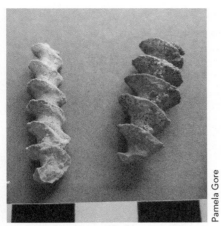

E. Screwlike central axis of the fossil bryozoan *Archimedes* (Mississippian). (Scale in centimeters.)

F. Screwlike central axis of the bryozoan *Archimedes* (Mississippian) embedded in limestone. Note the traces of lacy lattice structure surrounding the central axis. (Scale in centimeters.)

G. Unusually well preserved *Archimedes* bryozoan in which the central axis is completely surrounded by the lacy fenestrate lattice structure. (Scale in centimeters.)

H. Rounded cobble containing remains of branching bryozoans, Guelph, Ontario, Canada. The cobble is about 8 cm across.

Figure 10.7 Bryozoans.

I. Lacy network of fenestrate bryozoan. The word *fenestrate* means "window," referring in the lacy, screenlike lattice structure.

J. Fossil bryozoan, *Constellaria*, named for star-shaped patterns on its surface. (Scale in centimeters.)

IV. Brachiopoda: (Phylum Brachiopoda)

There are two groups of brachiopods: inarticulate and articulate.

Name
Brachio means "arm," and *pod* means "foot."

Chief Characteristics
Bivalved (two shells), each with bilateral symmetry. The plane of symmetry passes through the center of each shell or valve.

Geologic Range
Lower Cambrian to Recent (Quaternary).

Mode of Life
Brachiopods inhabit shallow marine environments; they generally live attached to the sea floor by a fleshy **pedicle**.

Inarticulate Brachiopods (Class Inarticulata) (Figure 10.8)
Inarticulate brachiopods have two valves of similar size and shape that are held together by soft tissue, with a fleshy pedicle between them. The word *inarticulate* means that there is no hinge holding the valves together. The valves are composed of apatite (calcium phosphate mineral) or an organic material called chitin. *Lingula* is a well-known inarticulate brachiopod.

A. Modern inarticulate brachiopod, *Lingula*. Note the fleshy pedicle. (Scale in centimeters.)

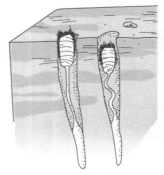

B. Inarticulate brachiopods in living position in a burrow within the sediment. The pedicle is attached at the bottom. (From Levin, H., 2013, *The Earth Through Time* (10th edition), figure 12-28c, p. 351. This material is reproduced with permission of John Wiley & Sons, Inc.)

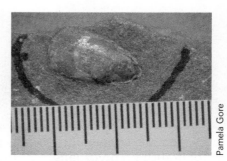

C. Fossil inarticulate brachiopod (Cambrian). (Scale in millimeters.)

Figure 10.8   Inarticulate Brachiopods.

Articulate brachiopods (Class Articulata) (Figure 10.9)

Articulate brachiopods have two valves attached together at a hinge line. The valves differ in size and shape and are composed of calcite. The larger valve has a circular opening near the hinge line through which the pedicle extended in life. Common articulate brachiopods include *Pentamerus, Rafinesquina, Atrypa, Leptaena, Platystrophia*, and *Spirifer*.

A. Modern articulate brachiopod. Note the circular opening (at top) through which the pedicle extended in life. (Scale in centimeters.)

B. Articulate brachiopods in living position. They attach to the sea floor with the pedicle. (From Levin, H., 2013, *The Earth Through Time* (10th edition), figure 12-28a, p. 351. This material is reproduced with permission of John Wiley & Sons, Inc.)

C. Fossil brachiopods. Pencil with scale in centimeters.

D. Articulate brachiopods in limestone, *Spirifer* (Mississippian to Lower Pennsylvanian).

Figure 10.9 Articulate brachiopods.

E. Fossil articulate brachiopods, *Platystrophia ponderosa* (Ordovician), McMillan Formation, Elk Creek, Kentucky. (Scale in centimeters.)

F. Fossil articulate brachiopods, *Spirifer*. (Scale in centimeters.)

V. Mollusca: Clams, Oysters, Snails, Slugs, Squid, Octopus, Cuttlefish (Phylum Mollusca) (Figure 10.10–10.18)

Name

Mollusca means "soft bodied."

Chief Characteristics

The molluscs are a diverse group of soft-bodied organisms; many have shells or other hard parts.

Composition

Aragonite, calcite, or a combination of both minerals.

Geologic Range

Cambrian to Recent (Quaternary).

Mode of Life

Molluscs inhabit marine, freshwater, and terrestrial environments. Some swim, some float or drift, some burrow into mud or sand, some bore into wood or rock, some attach themselves to rocks, and some crawl.

A. Bivalvia: Clams, Oysters, Scallops, Mussels, Rudists (Class Bivalvia) (formerly Pelecypoda) (Figure 10.10)

Name
Bivalvia means "two shells": *bi* means "two" and *valvia* means "shells."

Chief Characteristics
Skeleton consists of two calcareous **valves** (or shells) connected by a dorsal hinge. The exterior of the shell typically has radial or concentric ornamentation (growth lines), and the interior typically has muscle scars where the soft parts were attached to the shell. Shells are typically mirror images of each other, with the plane of symmetry passing between the two valves. Oysters and scallops are the exception: they have valves of different shape and size.

Composition
Most are aragonite, but oysters and scallops have calcite shells.

Geologic Range
Early Cambrian to Recent (Quaternary.

Mode of Life
Marine and freshwater. Many species are **infaunal** (burrowers living within the sediment) or borers (drilling into hard substrates), and others are **epifaunal** (living on the sediment surface).

A. Fossil bivalves, *Mercenaria (left)* and *Isocardia (right)* (Miocene), Maryland.

B. Modern bivalve molluscs, Atlantic coast. (Scale in centimeters.)

C. Fossil oyster, *Exogyra*, from Mississippi.

Figure 10.10 Bivalve molluscs.

D. Fossil scallop, *Chesapecten* (Miocene), Calvert Cliffs, Lusby, Maryland. (Scale in centimeters.)

B. Gastropoda: Snails and Slugs (Class Gastropoda) (Figure 10.11)

Name

Gastro means "stomach" and *pod* means "foot."

Chief Characteristics

Asymmetrical, spiral-coiled shell.

Composition

Most are aragonite, but some are calcite. Slugs do not have a shell.

Geologic Range

Early Cambrian to Recent (Quaternary).

Mode of Life

Marine, freshwater, or terrestrial.

A. Land snail, showing soft parts.

B. Fossil moon snail (Miocene), Maryland.

C. Fossil *Turritella* gastropods. Shells are about 6 cm long.

D. Modern gastropod shell. (Scale in centimeters.)

Figure 10.11   *Gastropods.*

C. Cephalopoda: Squid, Octopus, *Nautilus*, Cuttlefish (Class Cephalopoda)

Name

Kephale means "head," and *pod* means "foot."

Chief Characteristics

Symmetrical cone-shaped shell with internal partitions called **septae** (singular, septum). The shell may be straight or coiled in a spiral that lies in a plane. **Sutures**, visible on the outside of some fossils, mark the place where the septae join the outer shell. Cephalopods include the following groups: Nautiloidea, Ammonoidea, and Coleoidea (belemnites, cuttlefish, squid and octopus).

Composition

Aragonite.

Geologic Range

Late Cambrian to Recent (Quaternary).

Mode of Life

Marine only. Cephalopods are carnivorous (meat-eating) swimmers.

1. Nautiloids (Subclass Nautiloidea) (Figure 10.12)

Nautiloid cephalopods have straight or coiled shells with smoothly curved septa that produce simple straight or gently curved sutures. Their geologic range is Cambrian to Recent (Quaternary).

A. Model of living *Nautilus*.

B. Modern *Nautilus* shell. (Scale in centimeters.)

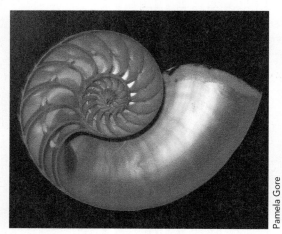

C. Interior of modern *Nautilus* shell showing spiral form, curved partitions (septae), and larger living chamber.

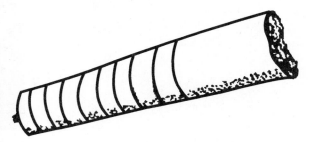

D. Sketch of a straight cone nautiloid cephalopod fossil showing the gently curved sutures on the outside.

E. Polished section through a straight cone nautiloid *Orthoceras*. (Ordovician), Morocco. (Scale in centimeters.)

Figure 10.12   Nautiloid cephalopods.

2. Ammonoids (Subclass Ammonoidea) (Figure 10.13)

Ammonoid cephalopods have straight or coiled shells with complexly folded septae that produce angular, wrinkled, or dendritic sutures. Their geologic range is Devonian to Cretaceous; all ammonoids are extinct. There are three basic types of sutures in ammonoid shells:

Goniatitic: relatively simple undulations

Ceratitic: smooth "hills" alternating with saw-toothed "valleys"

Ammonitic: complexly branching and dendritic or treelike

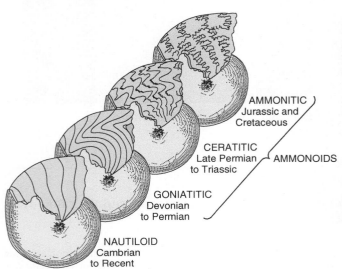

B. Goniatitic suture pattern on an ammonoid.

A. Comparison between the suture patterns of nautiloid and ammonoid cephalopods, showing the three basic types of ammonoid sutures.
(From Levin, H., 2013, *The Earth Through Time* (10th edition), figure 12-39, p. 355. This material is reproduced with permission of John Wiley & Sons, Inc.)

C. Ammonitic suture pattern on an ammonoid, *Manticoceras*, Rosebud County, Oklahoma.

D. Ammonoid that has been cut and polished to show the internal structure of septae. The ammonitic suture pattern is visible on the outside of the shell. (Scale in centimeters.)

Figure 10.13 Ammonoid cephalopods (*Continued*).

E. Fossil ammonoids in matrix. (Hand for scale (at top).)

F. Fossil ammonoid with ammonitic suture pattern.

Figure 10.13 Ammonoid cephalopods.

3. Coleoidea: Belemnites, Squid, Cuttlefish, Octopus, and their Relatives (Subclass Coleoidea)

a. Belemnoidea (Belemnites) (Figure 10.14)

Most **belemnites** have an internal calcareous shell, called a **rostrum**, that resembles a cigar in size, shape, and color. The front part of this shell is chambered, similar to nautiloids and ammonoids. The rostrum is made of fibrous calcite, arranged in concentric layers. (The rostrum is not present in some genera, such as *Acanthoteuthis*.) Their geologic range is Mississippian to Eocene; all are extinct.

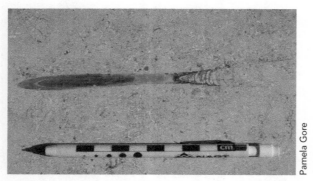

A. Fossil belemnite, *Hibolites* in the Solnhofen Limestone (Jurassic), Germany. (Pencil with centimeters for scale.)

B. Fossil belemnites. (Pencil with centimeters for scale.)

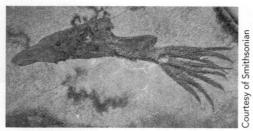

C. Soft part preservation of *Acanthoteuthis,* a type of belemnite, related to the squid (Lower Jurassic), Germany.

Figure 10.14 Belemnites.

b. Neocoleoidea: Octopus, Squid, Cuttlefish, and their Relatives
i. Sepiida: Cuttlefish (Order Sepiida) (Figure 10.15)

The cuttlefish is a swimming cephalopod mollusc, not a fish. It has a broad, white internal calcareous shell composed of aragonite, which is sold in pet stores as a calcium supplement for pet birds. Its geologic range is Jurassic to Recent (Quaternary).

A. Living cuttlefish.

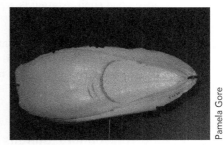

B. Cuttlebone, the internal shell of a cuttlefish.

Figure 10.15   Cuttlefish (Order sepiida)

ii. Teuthida: Squid (Order Teuthida)

Squid have eight arms, arranged in pairs, and two longer tentacles. There is a feather-shaped internal hard part, called a pen, made of an organic chitin-like material. The geologic range is Jurassic to Recent (Quaternary).

iii. Octopoda: Octopus (Order Octopoda) (Figure 10.16)

The octopus has eight arms. It has no hard parts, and soft body preservation is rare. The geologic range is Cretaceous to Recent (Quaternary).

Figure 10.16   Living octopus.

D. Scaphopoda: Tusk Shells (Class Scaphopoda) (Figure 10.17)

Chief Characteristics
Curved tubular shells open at both ends.

Geologic Range
Ordovician to Recent (Quaternary).

Mode of Life
Marine.

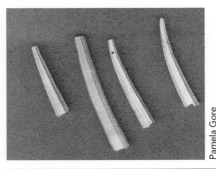

Figure 10.17   Scaphopods, *Dentalium sexangulare* (Pliocene). Piacenza Blue Clay, Castell'Arquato, Italy. About 5 cm long.

E. Monoplacophora

Name
Monoplacophora means "one plate."

Chief Characteristics
Cap-shaped shell. Animal is segmented and bilaterally symmetrical.

Geologic Range
Cambrian to Recent (Quaternary). Fairly common in certain lower Paleozoic strata; not known after Devonian until the Recent. *Neopilinia*, dredged from deep water near Central America, is one of the best examples of a "living fossil."

Mode of Life
Marine.

F. Polyplacophora: Chitons or Amphineurans (Class Polyplacophora) (Figure 10.18)

Name
Polyplacophora means "many plates."

Chief Characteristics
Shell consists of eight separate calcareous plates.

Geologic Range
Cambrian to Recent (Quaternary). Rare in fossil record; represented mainly by isolated plates.

Mode of Life
Marine. Chitons live on rocks or hard substrates in the surf zone.

Pamela Gore

Figure 10.18   Modern chiton, a type of mollusc. (Scale in centimeters.)

VI. Arthropoda: Insects, Spiders, Shrimp, Crabs, Lobsters, Barnacles, Ostracodes, Trilobites, Eurypterids (Phylum Arthropoda)

In lab, you will see examples of trilobite fossils, and you might see examples of some of the other groups.

Name
Arthro means "jointed" and *pod* means "foot."

Chief Characteristics
Segmented body with a hard external covering called an **exoskeleton**. Jointed legs.

Composition
Most fossil groups are calcium carbonate or calcium phosphate. Some are organic material only.

Geologic Range
Cambrian to Recent (Quaternary).

Mode of Life
Arthropods inhabit a wide range of environments, and their fossils have been found in both marine and freshwater sedimentary rocks.

A. Trilobita: Trilobites (Class Trilobita) (Figure 10.19)

Name
Tri means "three" and *lobus* means "lobed."

Chief Characteristics
Body has a three-lobed appearance: Two long grooves running from the head to the tail divide the body into three lobes. The trilobite body is also divided into three segments, from front to back:

Cephalon: Rigid head segment, commonly with compound eyes

Thorax: Jointed, segmented, flexible middle section

Pygidium: Rigid tail piece

Geologic Range
Cambrian to Permian. All are extinct.

Mode of Life
Exclusively marine.

A. Trilobite *Phacops rana* (Devonian), Silica Shale, Ohio.

B. Trilobite from the Conasauga Group (Cambrian), Northwest Georgia. About 3 cm long.

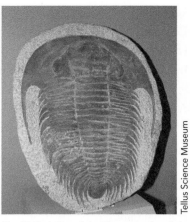

C. Trilobite *Acadoparadoxides briareus* (Cambrian), Morocco.

D. Trilobite *Asaphus* (Ordovician), Russia.

E. Trilobite *Dikelocephalus* (Cambrian), Morocco.

F. Unusual spiny trilobite, Morocco. (About 8 cm long.)

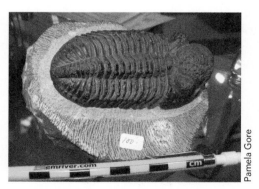

G. Trilobite. (Scale in centimeters.)

Figure 10.19 Trilobites.

B. Chelicerata: Horseshoe Crabs, Scorpions, Spiders, Mites, Eurypterids, and their Relatives (Subphylum Chelicerata)

1. Merostomata: Horseshoe Crabs and Eurypterids (Class Merostomata) (Figure 10.20)

a. Xiphosura: Horseshoe Crabs (Order Xiphosura)

Geologic range is Silurian to Recent (Quaternary).

A. Modern horseshoe crab, *Limulus polyphemus,* Jekyll Island, Georgia. (About 45 cm long).

B. Fossil horseshoe crab, Solnhofen Limestone (Jurassic), Germany.

Figure 10.20 Horseshoe crabs.

b. Eurypterida: Eurypterids (Order Eurypterida) (Figure 10.21)

The eurypterids are extinct scorpion-like or lobster-like arthropods that for a time were the dominant predators in the Paleozoic seas. They have a semicircular head with compound eyes, and a segmented rear section ending in a long projection called a **telson**. Two swimming legs are prominent near the head. Their geologic range is Ordovician to Permian; all are extinct. They were most abundant in the Silurian and Devonian.

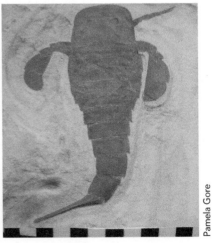
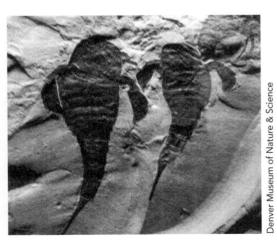

A. Eurypterid. (Scale in centimeters).

B. *Eurypterus remipes* (Early Silurian, 425 million years old), Fiddler's Green Formation, New York. About 27 cm long.

Figure 10.21 Eurypterids.

2. Arachnida: Scorpions, Spiders, Ticks, and Mites (Class Arachnida) (Figure 10.22)

Chief Characteristics
The Arachnids have eight legs.

Geologic Range
Devonian to Recent (Quaternary).

Florissant Fossil Beds National Monument

Figure 10.22   Unidentified spider fossil, Florissant Formation (Late Eocene), Colorado. (Scale in millimeters.)

C. Crustacea: Shrimp, Crabs, Lobsters, Crayfish, Barnacles, and Ostracodes (Subphylum Crustacea)

1. Ostracoda: Ostracodes (Class Ostracoda) (Figure 10.23)
Ostracodes consist of a tiny bivalved calcite shell encasing a shrimp-like creature. (See Laboratory 9: Microfossils and Introduction to the Tree of Life). Their geologic range is Cambrian to Recent (Quaternary).

Pamela Gore

Figure 10.23   Ostracode fossils (*Darwinula*) in brown shale (Triassic), Durham, North Carolina. (Scale in millimeters.)

2. Cirripedia: Barnacles (Class Maxillopoda) (Figure 10.24)
The barnacles live attached, and consist of calcareous (calcite) plates surrounding a shrimplike body. Their geologic range is Silurian to Recent (Quaternary).

Figure 10.24   Fossil barnacles, *Balanus concavus,* Miocene, Chesapeake Bay, Maryland. These barnacles are encrusting a fossil oyster shell. (Scale in centimeters.)

3. Malacostraca: Lobsters, Shrimp, Crabs, Crayfish, Sowbugs (Class Malacostraca) (Figure 10.25)

a. Decapoda: Lobsters, Shrimp, Crabs, Crayfish (Order Decapoda)

The name *decapod* comes from *deca* meaning "ten" and *pod* meaning "feet." These organisms have ten legs. Most are marine, but some inhabit freshwater, and a few are terrestrial. Their geologic range is Permian to Recent (Quaternary).

A. Fossil crab (*Avitelmessus*) from the Ripley Formation, (Cretaceous), Union County, Mississippi.

B. Fossil shrimp from the Solnhofen Limestone (Jurassic), Germany.
(Pencil with centimeters for scale.)

Figure 10.25   Fossil malacostracans.

b. Isopoda: Sowbugs, Pillbugs (Order Isopoda)

The name *isopod* comes from *iso* meaning "same" and *pod* meaning "foot," referring to their seven pairs of legs that are similar in size and shape.

The isopods are marine, freshwater or terrestrial. Segmented, gray pill bugs about 1 cm long live on land under rocks or wood. They can curl up into a ball when touched.

Their geologic range is Carboniferous to Recent (Quaternary).

4. Branchiopoda: Brine Shrimp, Clam Shrimp, and Their Relatives (Class Branchiopoda)

a. Conchostraca: Clam Shrimp (Order Conchostraca) (Figure 10.26)

Conchostracans, commonly known as clam shrimp, have weakly calcified, organic, clamlike shells with prominent growth lines. They primarily inhabit freshwater. The eggs can withstand drying and they are sometimes found in vernal pools on granite outcrops. Their geologic range is Devonian to Recent (Quaternary).

Figure 10.26 Conchostracans (*Cyzicus*) from the Triassic–Jurassic Newark Supergroup, North Carolina and Virginia. (Scale in millimeters at *left*. At *right*, bar scale is 1 millimeter.)

b. Notostraca: Tadpole Shrimp, *Triops* (Order Notostraca) (Figure 10.27)

Chief Characteristics

Notostracans have a circular shield with a triangular opening at the back, over a segmented body with a forked tail. They inhabit temporary pools and shallow freshwater to brackish or saline lakes. *Triops* eggs are commonly sold to hatch and grow in an aquarium as prehistoric sea monsters, as their geologic record goes back 220 million years.

The geologic range of notostracans is Carboniferous to Recent (Quaternary).

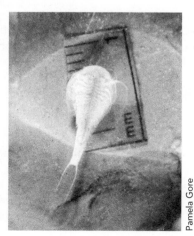

Figure 10.27 Notostracan fossils (genus *Triops*) from the Culpeper Basin, Newark Supergroup (Triassic), Manassas, Virginia. Specimen at right is modern *Triops* next to a fossil *Triops*. (Scale in millimeters.)

c. Cladocera: Water Fleas (Order Cladocera)

Cladocerans or water fleas mostly inhabit freshwater inland aquatic environments; a few are marine. Their geologic range is Permian to Recent (Quaternary).

d. Anostraca: Fairy Shrimp, Brine Shrimp or "Sea Monkeys" (Order Anostraca)

Fairy shrimp or brine shrimp inhabit lakes with a broad range of salinity, including hypersaline lakes (with higher salinity than sea water). Their eggs withstand drying and are commonly sold to hatch and grow in an aquarium as "sea monkeys". Their geologic range is Upper Cambrian to Recent (Quaternary).

D. Hexapoda (Subphylum Hexapoda)

1. Insecta: Insects (Class Insecta) (Figure 10.28)

Chief Characteristics

The insects are among the most diverse living group on Earth, although they are rarely found as fossils. The body is divided into three parts: head, thorax, and abdomen. The thorax bears six legs. Wings may be present or absent. Hard parts are organic material.

Geologic Range

Middle Devonian to Recent (Quaternary).

A. Wasp (*Palaeovespa*) (FLFO 51). Florissant Formation (Late Eocene), Colorado. (Scale in millimeters.)

B. Robber fly (FLFO 1020A). Florissant Formation (Late Eocene), Colorado. (Scale in millimeters.).

C. Unidentified Diptera (FLFO 7092A). Florissant Formation (Late Eocene), Colorado. (Scale in millimeters.)

D. Fossil dragonfly. (Pencil with centimeters for scale.)

E. Unidentified fossil insect. (Scale in millimeters.)

Figure 10.28   Insect fossils.

VII. Echinodermata: Starfish, Sea Urchins, Sand Dollars, Crinoids, Blastoids (Phylum Echinodermata)

Name
Echinos means "spiny" and *derma* means "skin."

Chief Characteristics
Five-part symmetry (derived from bilateral symmetry of the larval stages.) The five-rayed design is best displayed by the series of five **ambulacral** areas with pores, where the tube feet of the **water vascular system** protrude. **Tube feet** function for respiration, locomotion, attachment, and moving food toward the mouth. In some groups of echinoderms, the body is covered by a mosaic of small calcite plates, many of which have spines attached. In other groups, calcite plates or irregularly-shaped spicules are embedded in the tissues.

Composition
Calcite

Geologic Range
Cambrian to Recent (Quaternary).

Mode of Life
The echinoderms are exclusively marine. Some are attached to the sea floor by a stem with roots; others are free-moving bottom dwellers.

A. Crinoidea: Crinoids or "Sea Lilies" (Class Crinoidea) (Figure 10.29)
Crinoids are animals that resemble flowers. They consist of a **calyx** with **arms**, which is attached to the sea floor by a **stem** of calcite disks called **columnals**. Some living crinoids are free swimmers.

Geologic Range
Ordovician to Recent (Quaternary).

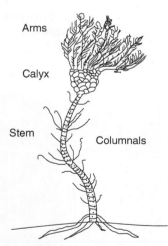

A. Diagram of the parts of a crinoid.

B. Crinoid showing rare fine details of calyx and arms.

C. Crinoids, showing calyx, arms and stem.

D. Silicified crinoid columnals from the Fort Payne Chert (Mississippian), Summerville, Georgia. (Scale in centimeters.)

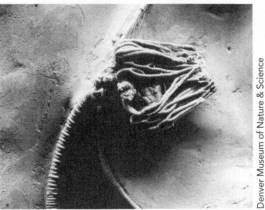

E. Crinoid, Edwardsville Formation (Early Mississippian), Indiana.

Figure 10.29   Crinoids.

B. Blastoidea: Blastoids (Class Blastoidea) (Figure 10.30)

Chief Characteristics
Blastoids are extinct animals with an armless budlike calyx on a stem.

Geologic Range
Ordovician to Permian. All are extinct.

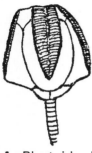

A. Blastoids showing five-part symmetry.   **B.** Blastoid calyx, *Pentrimites*. (Scale in centimeters.)

Figure 10.30   Blastoids.

C. Asteroidea: Starfish (Class Asteroidea) (Figure 10.31)

Name	*Asteroidea* means "starlike."
Chief Characteristics	Starfish are star-shaped echinoderms with five arms.
Geologic Range	Ordovician to Recent (Quaternary).

A. Living starfish showing ambulacral grooves with tiny white suction cup–like tube feet.   **B.** Dried modern starfish. (Scale in centimeters.)

C. Dried modern starfish. (Scale in centimeters.)

Figure 10.31   Starfish.

D. Ophiuroidea: Brittle Stars (Class Ophiuroidea) (Figure 10.32)

Chief Characteristics
Brittle stars have five arms, but they are very thin and flexible compared to those of the Asteroidea.

Geologic Range
Early Ordovician to Recent (Quaternary).

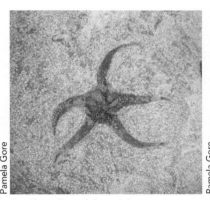

A. Fossil ophiuroids (Cambrian), Morocco.   **B.** Fossil ophiuroids.

C. Fossil ophiuroid. (Ordovician), Morocco.

Figure 10.32   Fossil ophiuroids.

E. Echinoidea: Sand Dollars and Sea Urchins (Class Echinoidea) (Figure 10.33)

Chief Characteristics
Echinoids are disk-shaped, biscuit-shaped, or globular. They range from circular to somewhat irregular in shape, but with five-part symmetry. In life, echinoids are covered with spines that attach to the rounded bumps covering their surfaces. Typically, the mouth is on the bottom and the anus is on top.

The five-rayed pattern on the sand dollars and sea urchins is their ambulacral areas, which have pores where the tube feet of the water vascular system protrude through the shell. Tube feet function for respiration, locomotion, attachment, and moving food toward the mouth.

Geologic Range
Ordovician to Recent (Quaternary).

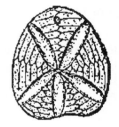

A. Echinoids.

B. Living echinoids or sea urchins. The five-part symmetry is visible through the spines.

C. Dried echinoid, showing several sizes of spines. (Scale in centimeters.)

D. Modern echinoid or sea urchin without its spines. (10 cm in diameter.)

E. Modern echinoid or sea urchin without its spines. (6 cm in diameter.)

F. Modern sand dollars. The one on the left is covered with tiny spines. (Scale in centimeters.)

Figure 10.33   Echinoidea (sand dollars and sea urchins) (*Continued*).

G. Modern echinoid called a sea biscuit. (Scale in centimeters.)

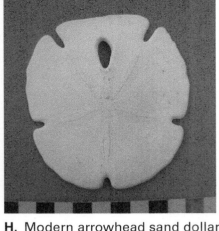

H. Modern arrowhead sand dollar. (Scale in centimeters.)

I. Fossil echinoid *Eupatagus antillarum.* (Eocene). (About 5 cm wide.)

J. Fossil echinoid *Periarchus quinquefarius* from the Sandersville Limestone (Eocene), Sandersville, Georgia. (6 cm wide.)

K. Fossil sand dollars, *Scutella subrotunda*, (Miocene) near Bordeaux, France.

L. Fossil sand dollars, *Periarchus lyelli,* from the Tivola Limestone (Eocene), Clinchfield, Georgia.

Figure 10.33   Echinoidea (sand dollars and sea urchins).

F. Holothuroidea: Sea Cucumbers (Class Holothuroidea) (Figure 10.34)

Chief Characteristics
Soft bodied with microscopic hard parts in the body wall.

Geologic Range
Middle Cambrian? or Middle Ordovician to Recent (Quaternary).

Figure 10.34 Living holothuroid. Note the tiny light-colored tube feet on the bottom.

VIII. Hemichordata: Graptolites (Phylum Hemichordata)

A. Graptolithina: Graptolites (Class Graptolithina) (Figure 10.35)

Name
Grapto means "write" and *lithos* means "stone." Graptolites sometimes resemble pencil marks on the rock.

Chief Characteristics
Organic skeletons consisting of rows or lines of small tubes or cups. Tubes or cups branch off a main cord or tube called a **nema**. They can consist of one, two, or many branches. Most are flattened in black shales and mudstones.

Composition
Organic. Typically preserved as carbon films.

Geologic Range
Cambrian to Pennsylvanian; all are extinct. They were most abundant in the Ordovician and Silurian.

Mode of Life
Planktonic (colonies attached to floats), some were benthic.

10.35 Graptolite fossils.

SELECTED REFERENCES

Adl SM, Simpson AG, Farmer MA, et al. The new higher level classification of eukaryotes with emphasis on the taxonomy of protists. Journal of Eukaryotic Microbiology 52(5):399–451, 2005; http://onlinelibrary.wiley.com/doi/10.1111/j.1550-7408.2005.00053.x/abstract.

Burki F, Shalchian-Tabrizi K, Minge M, et al. Phylogenomics reshuffles the eukaryotic supergroups. PloS ONE 2(8):e790, 2007; http://www.plosone.org/article/info:doi%2F10.1371%2Fjournal.pone.0000790

Dunn CW, Hejnol A, Matus DQ, et al. Broad phylogenomic sampling improves resolution of the animal tree of life. Nature 452(7188): 745–749, 2008. http://www.nature.com/nature/journal/v452/n7188/full/nature06614.html

Science. A Tree of Life (2003). http://www.sciencemag.org/feature/data/tol/;

Tree of Life (Special Issue). Science 300(5626):1605–1832. http://www.sciencemag.org/content/300/5626.toc#SpecialIssue

Tree of Life web project page on Eukaryotes http://tolweb.org/tree

Invertebrate Macrofossils and Classification of Organisms

PRE-LAB EXERCISES

1. Invertebrate macrofossils are useful for biostratigraphic correlation and determining the ages of sedimentary rocks. In the **Geologic Range Table** below, indicate the geologic range of each invertebrate group by placing an x in the boxes corresponding to the geologic periods in which it lived. When you have completed the chart, you will be able to see which fossil groups coexisted in time, which fossil groups never coexisted, and times at which major extinction events occurred.

GEOLOGIC RANGE TABLE

Fossil Group	Paleozoic							Mesozoic			Cenozoic		
	€	O	S	D	M	ℙ	P	℞	J	K	℞	N	Q
Porifera													
Rugose corals													
Tabulate corals													
Scleractinian corals													
Bryozoans													
Brachiopods													
Bivalves													
Gastropods													
Ammonoids													
Nautiloids													
Belemnoids													
Trilobites													
Eurypterids													
Barnacles													
Insects													

Fossil Group	Paleozoic							Mesozoic			Cenozoic		
	€	O	S	D	M	ℙ	P	T͟	J	K	ℙ	N	Q
Ostracodes													
Crinoids													
Blastoids													
Echinoids													
Asteroids													
Graptolites													

Key: € = Cambrian, O = Ordovician, S = Silurian, D = Devonian, M = Mississippian, ℙ = Pennsylvanian, P = Permian, T͟ = Triassic, J = Jurassic, K = Cretaceous, ℙ = Paleogene, N = Neogene, Q = Quaternary.

Use the completed Geologic Range table to answer the following questions.

2. Which fossil groups were present during the Paleozoic?
   All except the _____

3. Which fossil groups were present during the Mesozoic? _____
   _____
   _____

4. Which fossil groups were present during the Cenozoic? _____
   _____
   _____

5. Major faunal extinctions occurred several times in the geologic past. The boundaries between eras were drawn where mass extinction events occurred. The earlier of these two extinctions occurred at the end of the _____ Period (not era), _____ million years ago. List five fossil groups that became extinct at this time:
   _____
   _____

6. The second of these two mass extinctions was most dramatic among the vertebrates (dinosaurs and marine reptiles), but it also affected microfossils and invertebrate macrofossils. This other major extinction occurred at the end of the _____ Period (not era), _____ million years ago. The following invertebrate macrofossil fossil group(s) became extinct at this time:
   _____.

7. If you found a rock that contained both scleractinian corals and ammonoids, to which geologic periods (not eras) might it belong?
   _____
   _____

LAB EXERCISES

1. Identify all fossil specimens provided by your instructor. Using the information contained in this laboratory, fill in the table below. Use the taxonomic classification listed in this lab, beginning with the group with a Roman numeral (former phylum) and proceeding to various subgroups. Be as specific as possible.

    Optional: If additional reference materials are available, your instructor will ask you to identify the **genus** (as closely as possible) and find the geologic range of each fossil.

#	Taxonomic Group			Optional	
	I, II, . . . (Phylum)	A, B, C, . . . (Subphylum or Class)	1, 2, 3; a, b, c (Class, Subclass or Order)	Genus	Geologic Range of Genus
Example	Mollusca	Cephalopoda	Ammonoidea	*Bactrites*	Ordovician to Permian
1					
2					
3					
4					
5					
6					
7					
8					
9					
10					
11					
12					
13					
14					
15					
16					

#	Taxonomic Group			Optional	
	I, II, . . . (Phylum)	A, B, C, . . . (Subphylum or Class)	1, 2, 3; a, b, c (Class, Subclass or Order)	Genus	Geologic Range of Genus
17					
18					
19					
20					
21					
22					
23					
24					
25					

2. Compare brachiopod and bivalve shells.

   a. In the spaces below, draw sketches of bivalves and brachiopods to show the differences in symmetry. Then draw the symmetry planes.

Bivalve shell with the hinge at the top	Two bivalve shells together, in edge view, with the hinge on the left
Brachiopod shell with the hinge at the top	**Two brachiopod shells together, in edge view, with the hinge on the left**

b. In which does the plane of symmetry cut through the center of the valves?

_____

c. In which are the valves of two different sizes? _____

d. In which does the plane of symmetry pass between the valves? _____

e. In which are the valves the same size, but mirror images of one another?

_____

3. Compare a solitary rugose coral with a colonial rugose coral. Carefully sketch each, showing as many details as you can.

Solitary rugose coral	Colonial rugose coral

4. How do colonial rugose corals differ from tabulate corals?

_____

5. How do colonial rugose corals differ from scleractinian corals?

_____

6. How do corals differ from bryozoans, considering that both of them commonly consist of colonies of many individuals?

_____

7. a. How are gastropods similar to ammonoids?

_____

b. How do gastropods and ammonoids differ? (List at least two ways.)

_____

c. How do ammonoids and nautiloids differ?

_____

8. a. How are blastoids similar to crinoids?

_____

b. How do blastoids and crinoids differ?

_____

9. Identify the two fossils in the photo below. (Scale in centimeters.)
    a. Fossil on the left _____
    b. Fossil on the right _____

OPTIONAL ACTIVITY

Visit a seafood market to obtain and/or examine fresh squid, octopus, bivalves of various types, crabs, crayfish, shrimp and other organisms covered in this laboratory, if available. Suggested activities include the following:

- Compare the arms and tentacles of the squid; find its ink sac and internal pen.
- How does the octopus compare with the squid?
- Look at the various types of bivalve molluscs (clams, mussels, oysters, etc.), see how their two valves are attached together, and examine the muscle scars on the inside of the shells.
- Examine the shells and legs and of the crustaceans (crabs, lobsters, crayfish, etc.).
- Summarize and illustrate your findings.

Fossil Preservation and Trace Fossils

WHAT ARE FOSSILS?

Fossils are the prehistoric remains or traces of life that have been preserved by natural causes in the Earth's crust. Fossils include both the *remains* of organisms (such as bones or shells), and the *traces* of organisms (such as tracks, trails, and burrows, called *trace fossils*). Many species known as fossils also have living representatives; in other words, just because it's a fossil doesn't mean it's extinct.

HOW ARE ORGANISMS PRESERVED AS FOSSILS?

Most organisms that lived in the past left no record of their existence. Fossil preservation is a rare occurrence. To become preserved as a fossil, an organism must:

Have preservable parts. Hard parts (bones, shells, teeth, wood) have a much better chance of being preserved than do soft parts (muscle, skin, internal organs).

Be buried by sediment. Burial protects the organism from decay.

Escape physical, chemical, and biological destruction after burial. The remains of organisms can be destroyed by burrowing (bioturbation), dissolution, metamorphism, or erosion.

Organisms do not all have an equal chance of being preserved. The organism must live in a suitable environment for preservation. In general, marine and transitional (shoreline) environments are more favorable for fossil preservation than are continental environments, because the rate of sediment deposition tends to be higher.

TYPES OF FOSSIL PRESERVATION

The remains of organisms may be fossilized in a variety of ways, including preservation of unaltered hard parts, chemical alteration of hard parts, imprints of hard parts in the sediment, markings in the sediment made by the activities of organisms, and the rare preservation of unaltered soft parts.

Preservation of Unaltered Hard Parts (Original Material)

The shells of invertebrates and single-celled organisms, and the bones and teeth of vertebrates, may be preserved unaltered. Shells of invertebrates and single-celled organisms are most commonly composed of aragonite, calcite, or silica. Arthropod exoskeletons are composed of **chitin**, an organic (carbon-based) material, along with calcite and calcium phosphate in the exoskeletons of the crustaceans (e.g., crabs and lobsters). Bones and teeth of vertebrates are composed of phosphate minerals.

Aragonite

Aragonite shells of bivalves, snails, or scleractinian corals may be preserved unaltered in Cenozoic deposits (Figure 11.1). They are generally dissolved or recrystallized in older deposits. This is because aragonite is more soluble than calcite and because aragonite is metastable, and in time, it *recrystallizes to calcite.* There are examples of aragonite shells as old as Mississippian.

A. Aragonite bivalve shells (Miocene). Maryland.

Figure 11.1 Invertebrates with aragonite shells.

B. Aragonite gastropod shell (Miocene). Maryland.

Calcite

Hard parts made of calcite, such as echinoderms (crinoids, blastoids, echinoids), foraminifera, and a few bivalves (scallops and oysters) may be preserved unaltered (Figure 11.2).

A. Fossil oyster with calcite hard parts. *Exogyra* (Cretaceous). Stewart County, Georgia.

Figure 11.2 Invertebrates with calcite shells.

B. Calcite hard parts of fossil sand dollars, *Periarchus lyelli*, and scallop shell (Eocene). Tivola Limestone, Clinchfield, Georgia.

Silica

Hard parts made of silica, such as the skeletons of diatoms, radiolarians, and some types of sponges, may be preserved unaltered in some deposits (Figure 11.3).

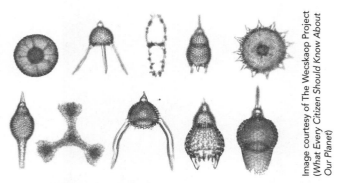

Figure 11.3 Hard parts of radiolarians are composed of silica.

Organic Material

Organic hard parts, or those made of carbon-based materials such as chitin, cellulose, keratin, sporopollenin, or collagen are present in some groups of organisms. Many arthropods, including the insects, have exoskeletons composed of chitin (an organic material similar in composition to our fingernails). Plant hard parts (wood) are composed of **cellulose**. Animal hair and fur are composed of **keratin** (Figure 11.4).

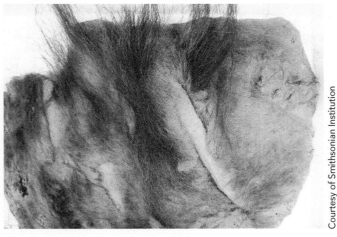

Figure 11.4 Preserved fur of a wooly mammoth (Pleistocene).

Phosphate

Hard parts made of phosphate minerals include the bones and teeth of vertebrates, conodonts, and the outer covering of trilobites (Figure 11.5). The shiny scales of some fossil fish are also phosphatic (Figure 11.6). Phosphate minerals include hydroxyapatite and calcium fluorapatite.

A. Fossil shark tooth, *Carcharodon megalodon.* (Scale in centimeters.)

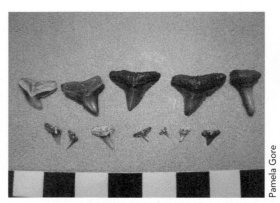

B. Fossil shark teeth. (Scale in centimeters.)

C. Mastodon teeth. (Scale in centimeters.)

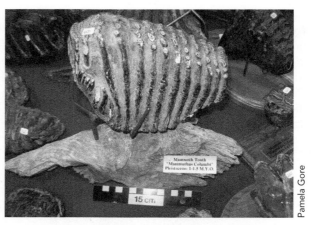

D. Wooly mammoth tooth, *Mammuthus columbi* (Pleistocene). (Scale in centimeters.)

Figure 11.5   Examples of hard parts (teeth) made of phosphate.

Figrue 11.6   Fossil fish scales and fins (Jurassic). Midland fish bed, Licking Run, Culpeper Basin, Midland, Virginia. (Scale in centimeters.)

Chemical Alteration of Hard Parts

The hard parts of many fossil organisms have been chemically altered by the addition, removal, or rearrangement of chemical constituents.

Permineralization

Permineralization is the filling of pores (tiny holes) in wood, shell, or bone by the precipitation or deposition of minerals from solution (Figure 11.7). The added mineral matter makes the permineralized fossil much heavier than the original material. Petrified wood is a common example of permineralization.

A. Detail of fossil whale bone showing pores that can become filled with mineral deposits (Miocene). Calvert Cliffs, Chesapeake Bay, Lusby, Maryland.

B. Petrified wood (*Araucarioxylon arizonicum*), found near Petrified Forest National Park, Arizona (Triassic).

Figure 11.7 Examples of Permineralization

Replacement

Replacement is the molecule-by-molecule substitution of another mineral of different composition for the original material (Figure 11.8). The fine details of shell structures are generally preserved. Minerals that commonly replace hard parts are silica and pyrite. Look for fossils that should be calcareous (crinoids, molluscs, brachiopods, corals) but that do not fizz in acid or that are golden metallic.

A. *Spirifer* brachiopods replaced by pyrite. (Devonian). The larger brachiopod is about 3 cm wide.

B. Brachiopod *Mucrospirifer* replaced by pyrite. Width of fossil is 5.5 cm.

Figure 11.8 Examples of hard parts replaced by pyrite or silica (*Continued*).

C. Pyritized ammonites (Jurassic), Germany.

D. Crinoid columnals replaced by silica, Fort Payne Chert, Georgia.

E. Pyritized ammonite, *Kosmoceras,* sliced in half to show internal structure, (Jurassic), Russia. (Scale in centimeters.)

Figure 11.8 Examples of hard parts replaced by pyrite or silica.

Recrystallization

Many modern mollusc shells are made of aragonite. Aragonite is a metastable form of calcium carbonate ($CaCO_3$). With time, the aragonite alters or recrystallizes to calcite, a stable form of $CaCO_3$ (Figure 11.9). Many Paleozoic shells that fizz in acid are probably recrystallized from the original aragonite to calcite (except for echinoderms and others that were originally calcite).

Figure 11.9 Recrystallized bivalve shells *Mercenaria permagna* (Late Pliocene to early Pleistocene), Fort Drum, Okeechobee County, Florida.

Carbonization

Carbonization (or *distillation*) preserves plants or animals as a thin carbon film, usually in fine-grained sediments (shales) (Figure 11.10). Fine details of the organisms may be preserved. Plant fossils, such as ferns and leaves, are preserved in shale by carbonization. Soft-bodied animals can also be preserved as carbonaceous films in black shales, for example, the Cambrian Burgess Shale fauna.

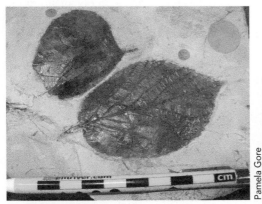

A. Carbonized leaves
(Scale in centimeters.)

B. Carbonized ferns, Pennsylvanian, Lookout Mountain, Georgia.

C. Carbonized angiosperm leaf.

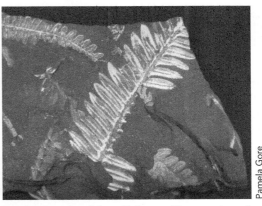

D. Carbonized fossil ferns (Pennsylvanian), St. Clair, Pennsylvania.

E. Carbonized fossil fish (Eocene), Green River Formation, Wyoming. The largest fish shown here is about 25 cm long.

F. Carbonized leaf and insect, Florissant Fossil Beds (Oligocene), Florissant, Colorado. (Scale in centimeters.)

Figure 11.10 Examples of carbonization of plants and animals.

Imprints of Hard Parts in Sediment

Many fossils are simply imprints with no shell material present at all. Hard parts are commonly destroyed by decay or dissolution after burial, but they can leave a record of their former presence in the surrounding sediment.

Impressions and Molds

Impressions or molds are the imprints of an organism (or part of an organism) in the sediment. A shell buried in sandstone may be leached or dissolved by groundwater, leaving a mold of the shell in the surrounding sandstone. There are two types of molds: external and internal.

External molds are imprints of the *outside* of a shell in the rock (Figure 11.11). If the original shell was *convex*, the external mold will be *concave*.

A. External mold of a bivalve, Morehead City, North Carolina. (Scale in centimeters.)

B. External mold of a crinoid stem in black shale, Maryland. (Scale in centimeters.)

Figure 11.11 Examples of external molds.

Internal molds are imprints of the *inside* of the shell in the rock (Figure 11.12). Look for such features as *muscle scars* that are present on the inside of bivalve shells. Internal molds are produced when a shell is filled with sediment that becomes cemented, and then the shell is dissolved away (Figure 11.13). Internal molds are sometimes called **steinkerns**.

A. Internal molds of several types of bivalves, showing muscle scars where soft parts attached to the inside of the shells, Morehead City, North Carolina. (Scale in centimeters.)

B. Internal mold of *Turritella* gastropods (Paleocene), Aquia Creek, Marlborough Point, Stafford County, Virginia. (Scale in centimeters.)

Figure 11.12 Examples of internal molds (*Continued*).

C. Internal mold of bivalve *Cucullaea gigantea* (Eocene), Largo, Maryland. (Scale in centimeters.)

D. Internal molds of *Turritella* gastropods.

Figure 11.12   Examples of internal molds.

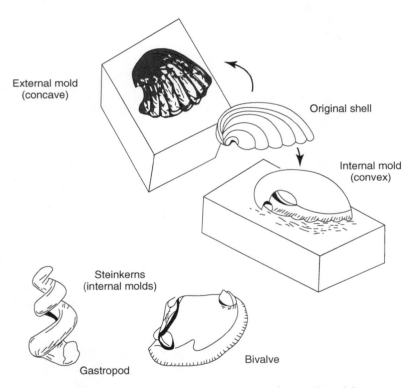

Figure 11.13   Formation of internal and external molds.

Casts

A cast may be produced if a mold is filled with sediment or mineral matter. A cast is a replica of the original. Casts are relatively uncommon. A rubber mold of a fossil can be filled with modeling clay or plaster of Paris to produce a replica or artificial cast of the original object (Box 11.1).

> **Box 11.1   Making a Mold or Cast**
>
> **Try this:**
>
> - You can make an external mold by pressing the outside of a shell into modeling clay.
> - You can make a cast by filling that mold with plaster of Paris.
> - You can make an internal mold by filling a shell with modeling clay.

Preservation of Unaltered Soft Parts

In rare circumstances, the soft parts of an organism are preserved (Figure 11.14). The two most common methods of preserving soft parts are **freezing** and **desiccation** (drying or mummification). For example, Pleistocene wooly mammoths are frozen in permafrost in Siberia and Alaska. Stories abound of frozen mammoth meat still being edible after about 10,000 years.

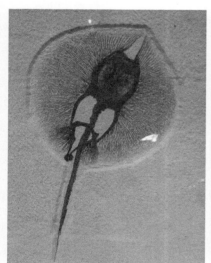
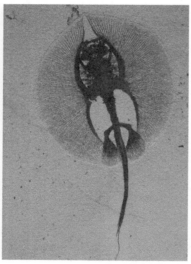

Figure 11.14   Soft part preservation of fossil stingrays (fish), *Heliobatis radians,* from the Green River Formation (Eocene), Kemmerer, Wyoming. The specimen on the *left* with claspers is male; a female is on the *right.* Specimens are about 30 cm long.

Soft parts of organisms such as insects or small frogs may be preserved if the organism becomes trapped in pine tree sap or resin (later altering to **amber**) (Figure 11.15).

Soft tissue of organisms can also be preserved in peat bogs such as those in northern Europe (e.g., Lindow Man, Tollund Man) because of the low temperatures, highly acidic water, and lack of oxygen.

A. Amber. Some with insects. (Scale in centimeters.)

B. Amber containing insect fossils. (Scale in centimeters.)

Figure 11.15   Insects preserved in amber.

Sometimes fossils are preserved in *more than one way*. For example, fossil fern leaves may be carbonized, but they may also leave impressions or external molds in the sediment. Also, organisms with porous skeletons may be preserved in several ways. A coral that was originally aragonite may be replaced by silica, or recrystallized to calcite, but at the same time, it may also have its original pore spaces filled by silica or calcite through the process of permineralization. Similarly, a bone with original material preserved also may also be permineralized if its pore spaces have been filled by minerals deposited from solution. Petrified wood is formed mainly through permineralization, but there may also be some replacement of the wood by silica, so it too can have more than one type of preservation.

Trace Fossils or Ichnofossils

Trace fossils are markings in the sediment made by the activities of organisms. They result from the movement of organisms across the sediment surface, from the tunneling of organisms into the sediment, or from the ingestion and excretion of sedimentary materials. The study of trace fossils is called **ichnology**. (See Laboratory 5: Sedimentary Structures.)

Trace fossils provide geologists with much useful information about ancient water depths, paleocurrents, availability of food, and rates of sediment deposition. In many cases, tracks of animals are the only record of their existence. For example, in many places, dinosaur tracks are much more abundant than dinosaur bones. During its lifetime, a single dinosaur makes millions of tracks, but it leaves only one skeleton that might or might not be preserved.

Tracks and Trackways

Tracks (footprints) are produced by the feet of a walking animal (Figure 11.16). A **trackway** is a continuous series of tracks made by one animal (Figure 11.17). From a study of trackways, the geologist can determine the leg length and height of an animal, whether it walked on two legs (bipedal) or four (quadrupedal), its speed and whether it was running or walking, and whether it was carnivorous (large claws), herbivorous (hooved), or aquatic (webbed feet).

Figure 11.16   Dinosaur tracks in shale (Triassic), Connecticut. (Tracks are about 15 cm long.)

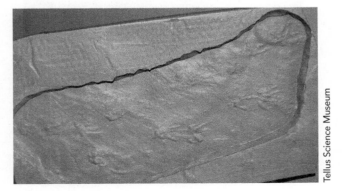

Figure 11.17   Cast of amphibian trackway (Pennsylvanian). Chatooga County, Georgia. (Specimen is about 50 cm high.)

Trails

Trails are produced as a worm or arthropod crawls or as the tail or belly of an animal drags the ground (Figure 11.18). Trails or *crawling traces* are usually linear and indicate movement in a particular direction. Some trails are more meandering in appearance and probably represent *grazing traces* as an invertebrate systematically combed an area of sediment for food. There are also *resting traces* produced by animals such as trilobites. Most trails are produced by organisms with *bilateral symmetry* that have a well-defined head (or anterior) and tail (or posterior) region.

A. *Climactichnites* trails in rippled sandstone (Late Cambrian), Quebec. Trails are about 10 to 20 cm wide.

Figure 11.18 Trails.

B. Trails in red siltstone (Triassic), Culpeper Basin, near Manassas, Virginia. (Scale in centimeters.)

Burrows

Burrows are the excavations of an animal that are made into *soft sediment* (Figure 11.19). Burrows are typically used as feeding and/or dwelling structures (Figure 11.20). Continued burrowing or **bioturbation** of the sediment destroys primary sedimentary structures, and results in a massive, homogeneous, structureless rock.

Figure 11.19 Branching burrows in siltstone (Triassic), Deep River Basin, North Carolina. Lens cap for scale is about 5 cm in diameter.

Figure 11.20 *Skolithos* worm burrows in quartzite stream cobbles (Cambrian), Henson Creek, Prince George's County, Maryland. (Scale in centimeters).

Borings

Borings are holes made by an animal into firm material, which could be shell, rock, wood, or hard cemented sediment (Figure 11.21). Borings are usually circular in cross section. Some snails produce borings or drill holes into other molluscs, such as clams. Some bivalves bore into wood, rock, or other hard materials. Sponges also produce borings, often riddling shells with numerous tiny holes.

A. Borings in bivalve shells, St. Augustine, Florida. (Largest shell is about 3 cm wide.)

B. Borings in fossil oyster produced by a clionid sulfur sponge. (Oyster is about 20 cm wide.)

Figure 11.21   Borings in bivalve shells.

Coprolites

Coprolites are the fossilized excrement of animals (Figure 11.22). Coprolites can contain fragments of undigested food that can provide valuable information about the feeding habits of fossil organisms. Coprolites smaller than 1 mm are called *fecal pellets*. Fecal pellets may be extraordinarily abundant in some environments, and they are the dominant *allochem* (or particle) in pelleted limestones. Some coprolites and fecal pellets have a high phosphate content. Coprolites have been mined for the phosphate, which is used for fertilizer.

A. Phosphate-rich coprolite from the Triassic Durham Basin of North Carolina. Note the rhombohedral fish scale embedded in the coprolite near the left end. (Scale in millimeters.)

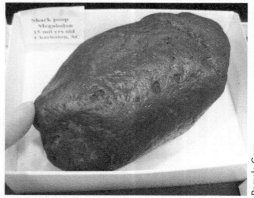

B. Phosphatic *Carcharadon megalodon* shark coprolite, Charleston, South Carolina. (Finger for scale.)

C. Coprolite. (Scale in centimeters.)

Figure 11.22

Gastroliths

Gastroliths are the highly polished stones from the gizzards of birds or from the stomachs of reptiles (including dinosaurs). Gastroliths or gizzard stones were probably used to grind food in the stomach of the animal.

Bite Marks

Bite marks of a predatory animal may be left on its prey (Figure 11.23).

Figure 11.23 V-shaped bite mark of a mosasaur on an ammonite, *Placenticeras meeki* (Late Cretaceous), Pierre shale, South Dakota.

Root Marks

Root marks are trace fossils produced by plants (Figure 11.24). They are present mainly in terrestrial sediments or soils, which are typically reddish brown. The root mark may be surrounded by green or gray zones where chemical reduction of iron has occurred as the root decayed. Root marks can bifurcate (or branch downward). In some areas, roots are also associated with **caliche** (calcite nodules in the soil) or sedimentary ironstones.

Figure 11.24   Root marks in siltstone (Triassic), Deep River Basin, near Sanford, North Carolina. (Rock hammer for scale.)

Stromatolites

Stromatolites are laminated, mound-like structures formed by colonies of sediment-trapping *cyanobacteria* (Figure 11.25). These organisms inhabit some carbonate tidal flats, and they produce dome-like laminations in lime mud (fine-grained limestone or micrite). Stromatolites are organo-sedimentary structures and not fossils because they contain no recognizable anatomical features. They can, however, be considered trace fossils.

A. Digitate (finger-like) stromatolites (Ordovician), western Maryland. (Finger for scale.)

B. Modern stromatolites, Shark Bay, Western Australia.

Figure 11.25   Stromatolites.

HOW LIKELY IS IT FOR AN ORGANISM TO BECOME PRESERVED AS A FOSSIL?

There are about 1.5 million known species of living plants and animals. In all, there may be as many as 4.5 million living species. In contrast, there are only about 250,000 known fossil species. The fossil record covers many hundreds of millions of years, and the living flora and fauna represent only one "instant" in geologic time. Thus, you might expect the number of fossil species to far outnumber the number of living species, if fossil preservation were a relatively common event. The fact that the number of fossil species is so small suggests that the preservation of organisms as fossils is extremely rare. It has been estimated that fewer than 10% of the animal species living today are likely to be preserved as fossils.

WHY IS PRESERVATION SO RARE?

One reason preservation is uncommon has to do with the environments that species inhabit. For example, the vast majority of living species are insects. (Of the 1.5 million known species, approximately 1 million are insects). Insects are rarely preserved as fossils because they generally live on dry land and are unlikely to be buried by sediment after death. If an organism is not rapidly buried after death, chances are it will be rapidly broken down by scavengers and bacterial decay.

Some groups of organisms that inhabit soft sediment in marine environments, such as molluscs, are much more likely to be preserved as fossils. In fact, for some groups of animals (brachiopods and cephalopods, for example), there are more fossil species than living species. This is particularly true for organisms that were much more abundant in the past and for groups that have suffered extinctions of many of their species.

Another reason that so few species are represented by fossils is that many organisms are soft-bodied and lack hard parts. For soft parts of organisms to be preserved, it is necessary to isolate them from oxygen almost immediately after death. This is most likely to occur when organisms are rapidly buried in fine-grained sediment in anoxic water, but this only happens in rare, isolated environments. There are a few spectacular examples of this rare type of preservation, including the Cambrian Burgess Shale of Canada. Soft-bodied forms are also preserved in the Late Precambrian Ediacara fauna of Australia. Feathers of the earliest birds, *Archaeopteryx*, are preserved in the Jurassic Solnhofen Limestone of Germany.

Fossil Preservation and Trace Fossils Exercises

PRE-LAB EXERCISES

1. The sketches below represent a cross section of a bivalve shell with the hinge at the left. Complete the drawings below by shading in the part that would become the part in the label.

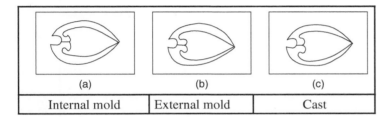

(a)	(b)	(c)
Internal mold	External mold	Cast

Hint #1: Remember the photo below from earlier in this lab (Fig, 11.12C).

15 cm

Pamela Gore

What type of preservation is this? _____

How does this fossil relate to the diagrams above? _____

Hint #2: Take two clam shells and hold them in an orientation corresponding to the sketches, and think about it.

Hint #3: Remember that an internal mold forms from the sediment that fills the space between the valves, and an external mold forms from the sediment *surrounding* the exterior of the valves. The cast forms in the area that the actual shell once occupied.

2. List the composition of the unaltered hard parts for each organism or part of an organism (for example, calcite, aragonite, phosphate, silica or organic).

Organism	Composition of Hard Parts
a. Shark tooth	
b. Clam	
c. Brachiopod	
d. Crinoid	
e. Scallop	
f. Whale bone	
g. Wood	
h. Snail	
i. Fish	
j. Sponge	
k. Diatom	
l. Foraminifera	
m. Coral	
n. Radiolarian	
o. Trilobite	

3. List the most common type(s) of preservation for the following fossils.

Fossil	Mode of Presentation
a. Petrified wood	
b. Fern	
c. Shark tooth	
d. Pyritized brachiopod	
e. Crinoid stem made of calcite	
f. Crinoid stem made of silica	
g. Dinosaur bone	
h. Calcite brachiopod	
i. Insect	
j. Bivalve	

4. Examine this photograph from Dinosaur Ridge near Morrison, Colorado, and answer the following questions.

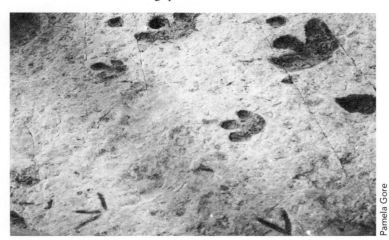

Pamela Gore

   a. Is the rock most probably of marine or nonmarine origin?

   _____

   b. What additional features would you look for to confirm your answer to question a, if you could examine the locality where these samples were collected? (List at least two things you would look for.)

   _____

   c. How many different animals left their mark on the rock? Explain.

   _____

   d. Are these tracks or trackways? _____

   e. What is a possible origin of the circular imprints near the larger three-toed tracks at the top of the image?

   _____

5. Examine the photograph below and answer the following questions.

Courtesy of Smithsonian Institution. Photo by Pamela Gore

   a. What type(s) of animal behavior are suggested by the trails?

   b. What types of marine invertebrates may have produced these trails called *Climactichnites*?

   _____

c. What other sedimentary structure is present here in addition to the trails?

_____

d. What type of rock is this most likely to be, because of this sedimentary structure?

_____

e. What was the most probable depositional environment for this slab of rock, based on your interpretation of the sedimentary structure?

_____

LAB EXERCISES

Making Molds and Casts

In this activity you will gain an appreciation for the differences between fossils, molds, and casts. Do this at the beginning of the lab so that the plaster of Paris will have time to set (takes about a half an hour).

Option 1

Materials required:
- Modeling clay or lightweight modeling compound such as Crayola® Model Magic®
- Bivalve shells from the beach or a seafood market or restaurant
- Plaster of Paris and water
- Small plastic dishes or disposable paper bowls for mixing plaster and water
- Plastic spoons for stirring plaster
- Measuring spoons (optional)

1. Take a bivalve shell and press it into modeling clay or modeling compound to make an external mold. Then carefully pull the shell out and examine the external mold.

2. Mix the plaster of Paris and water until it has the consistency of pancake batter. Pour it into the external mold you made in the modeling clay and let it harden. After the plaster hardens, remove it from the mold. This will be a cast.

3. Now take a bivalve shell and push modeling clay into the inside. Peel out the clay to see the internal mold.

Option 2

Materials required:
- Set of rubber fossil molds (such as Fossilworks Casting Kit)
- Plaster of Paris and water
- Small plastic dishes or disposable paper bowls for mixing plaster and water.
- Plastic spoons for stirring plaster
- Measuring spoons (optional)

1. Rubber molds are created from the original fossil. Using the rubber molds and plaster of Paris, make a cast (or replica) of the original fossil. Mix the plaster of Paris until it has the consistency of pancake batter and pour it into a rubber mold. When the plaster hardens, remove it from the mold. This will be a cast or replica of the original.

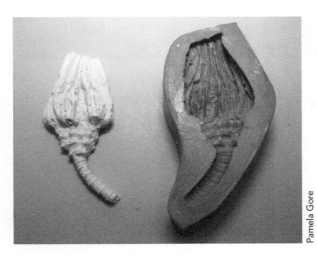

Pamela Gore

Plaster cast of fossil *(left)* and blue rubber mold *(right)*.

2. Examine the fossil specimens provided by your instructor. Identify the fossils, and give the type (or types) of preservation of each, as specifically as possible. If the hard parts are *original material*, specify the *composition* (calcite/aragonite, silica, phosphate). If the fossil is *chemically altered*, specify the *type of alteration* (replacement, permineralization, etc.) as well as the chemically altered *composition* (pyrite, silica, etc.). If you can see the imprints of hard parts in the sediment, specify whether they are *internal molds or external molds*. If you are looking at a trace fossil, specify the type.

Specimen	Type of Preservation*	Composition†	Identify the Fossil‡
1.			
2.			
3.			
4.			
5.			
6.			
7.			
8.			
9.			
10.			

Specimen	Type of Preservation*	Composition[†]	Identify the Fossil[‡]
11.			
12.			
13.			
14.			
15.			
16.			
17.			
18.			
19.			
20.			

*For example, internal mold, burrow, replacement, carbonization, unaltered hard part, coprolite, etc.
[†]For example, calcite, silica, pyrite, phosphate, sediment, etc.
[‡]For example, rugose coral, fern, articulate brachiopod, shark tooth, trilobite, etc.

3. Describe the preservation of the fossil specimen shown below.

Pamela Gore

_____

_____

4. Identify the specimen below as specifically as possible and describe the type(s) of preservation shown.

Pamela Gore

_____

_____

5. Explain why museums generally mount casts of dinosaur fossils for display, rather than the real specimens. Give three reasons, at least two of which are related to fossil preservation.

Tellus Science Museum

Cast of a _Tyrannosaurus rex_ nicknamed "Stan" from the Hell Creek Formation, (Cretaceous) South Dakota.

_____

_____

_____

6. Examine the specimen in the photograph below.

a. Explain the origin of the two circular openings on this specimen.

_____

b. Tell what sort of organism most likely made these circular openings, and why they are located in this particular place on the specimen.

_____

_____

c. What might have caused the light-colored circular patches near the top edge and on the left side of this specimen?

_____

Pamela Gore

Evolution of the Vertebrates

In this lab, you will visit several websites to learn about the evolution of the vertebrates. You will examine fossils at the University of California Museum of Paleontology website, the Tree of Life website, and others.

The lab consists entirely of research on the World Wide Web. Your instructor might direct you to do this as homework or as a project during lab time, if computers are available. Use the Web addresses to reach outside Web pages to look for the answers to the questions.

INTRODUCTION TO THE CHORDATA

Vertebrate Classification

The terminology used for classification of organisms changed in 2005, and some traditional groupings of organisms have been abandoned. Instead of Kingdoms, Phyla, Classes, and so on, there are now **nameless ranks**, with the ranked hierarchy indicated by indents or bullets. This approach has the advantage of flexibility and ease of modification with rapidly occurring advances in the study of gene sequencing. The traditional kingdoms of Animalia, Plantae, Fungi, and so on are now seen to be derived from lineages of single-celled organisms. (For your convenience, a simplified version of some of the traditional nomenclature is included.)

The classification of the vertebrates is shown in Box 12.1.

Box 12.1 Taxonomic Hierarchy

DOMAIN EUKARYOTA
Supergroup Opisthokonta

- **Animalia** (Metazoa or multicellular organisms) [Kingdom Animalia]
  - •• Bilateria (bilaterally-symmetrical animals)
    - ••• Deuterostomia (in the embryo, the first opening becomes the anus and the second opening becomes the mouth)
      - •••• **Chordata** [Phylum Chordata]
        - ••••• Cephalochordata (lancelets, amphioxis, *Pikaia*) [Subphylum Cephalochordata]
        - ••••• Urochordata (sea squirts) [Subphylum Tunicata]
        - ••••• Craniata (animals with skulls)
          - •••••• **Vertebrata** (animals with backbones) [Subphylum Vertebrata]
            - ••••••• Osteostraci (fossil armored jawless vertebrates known as ostracoderms) [Class Osteostraci]
            - ••••••• Euconodonta (conodonts)
            - ••••••• Gnathostomata (vertebrates with jaws) [Infraphylum Gnathostomata]

Fossil fish, *Knightia* (left), and stingray *Heliobatis radians*.
Eocene Green River Formation, Wyoming. (Pencil with centimeters for scale.)

Pamela Gore

Evolution of the Vertebrates Exercises

DEUTEROSTOMES

Use this reference to answer the questions below.

University of California Museum of Paleontology, Introduction to the Deuterostomia: http://www.ucmp.berkeley.edu/phyla/deuterostomia.html

1. Give an example of a deuterostome. _____

2. Most deuterostomes belong to one of two groups. What are they?

   _____ and _____

3. To what larger group do the deuterostomes belong? _____

University of California Museum of Paleontology, Introduction to the Chordata, http://www.ucmp.berkeley.edu/chordata/chordata.html.
Use this reference to answer the question below.

4. The vertebrates belong to the Chordata. What are four characteristics of the Chordata?

   _____

   _____

   _____

   _____

VERTEBRATES

Vertebrates are animals that have a dorsal hollow nerve cord surrounded by supportive material, the vertebral column (commonly called the spine or backbone), and an internal skeleton. The skeleton is usually composed of bone (hard material) and cartilage (soft and flexible material, such as that in your nose or ears), and serves for muscle attachment. In most vertebrates, the skeleton is first formed in cartilage and then replaced by bone before birth.

University of California Museum of Paleontology, Vertebrates: More on Morphology(2005), http://www.ucmp.berkeley.edu/vertebrates/vertmm.html.

Use this reference to answer the following questions.

5. The main bony disk-shaped or spool-shaped portion of the vertebra is called the

   _____.

6. The nerve or spinal cord passes through the _____.

7. Vertebrates possess two types of bone. What are they?

   _____

   _____

NONVERTEBRATE CHORDATES

Some chordates are not vertebrates and are not in the taxonomic group Craniata. Use these references to answer the questions below about nonvertebrate chordates.

Lundberg, J. G. (1995). Tree of Life Web Project, Chordata, http://www.tolweb.org/Chordata/. University of California Museum of Paleontology, Introduction to the Cephalochordata, http://www.ucmp.berkeley.edu/chordata/cephalo.html.

8. Name two nonvertebrate chordates.

    _____

    _____

9. The earliest fossil chordate is *Yunnanozoon lividum*.

    a. Where is it found? _____

    b. What is its age (Give period name.) _____

    c. What is its age in years? _____

Pikaia.

One of the earliest chordates is Pikaia, a pre-vertebrate with a notochord. See the notochord near the dorsal surface in the image above. The rib-like features are muscles, that give it a segmented appearance.

Hooper Virtual Paleontological Museum, (1996). Chordate, *Pikaia gracilens*, http://hannover.park.org/Canada/Museum/burgessshale/chordate.html.
Smithsonian Institution National Museum of Natural History, The Burgess Shale, http://paleobiology.si.edu/burgess/index.html.
Smithsonian Institution National Museum of Natural History, Burgess Shale, *Pikaia gracilens*, http://paleobiology.si.edu/burgess/pikaia.html.
Smithsonian Institution National Museum of Natural History, The Cambrian World, http://paleobiology.si.edu/burgess/cambrianWorld.html. (The painting of *Pikaia's* habitat in color gives a better illustration of its mode of life.)

10. Locate some basic facts about *Pikaia*.

    a. What was the mode of life of *Pikaia?* _____

    b. What is its age? (Give period name.) _____

    c   What is its age in years? _____

    d. In which well-known geologic formation is it found? _____

    e. To which group of chordates does it probably belong? _____

CONODONT ANIMAL

You learned about conodonts in the microfossil lab. For many years, scientists did not know what type of animal the conodonts came from. Use this reference to answer the questions below about the conodont animal.

Janvier, P. (1997). Tree of Life Web Project, Euconodonta, http://www.tolweb.org/Euconodonta/.

11. When and where was the conodont animal discovered?

    a. When _____

    b. Where _____

12. Describe what the conodont animal looked like.

    _____

13. How do we know that the conodont animal is a chordate? _____

14. Where in its body were the conodonts? _____

15. Why is it placed among the Craniata? _____

OSTRACODERMS: FOSSIL ARMORED JAWLESS FISH

The ostracoderms are extinct jawless fish that were covered in bony armor. The word *ostracoderm* means "bone skin": *ostra* means "bone" or "shell" and *derm* means "skin." Sometimes jawless fish are referred to as agnathans: *a* meaning "without" and *gnath* meaning "jaw." (Remember this by thinking of "gnashing of teeth" or "gnawing.") Ostracoderms were the dominant vertebrates in the Silurian and Early Devonian, but they declined after that time as the jawed vertebrates increased in diversity. The earliest fish fossils are found in marine rocks, suggesting that fish evolved in the sea.

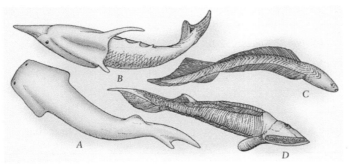

Early Paleozoic ostracoderms. These fish were 18–40 cm long. (From Levin, H., 2013, *The Earth Through Time* (10th edition), figure 12-63, p. 365. This material is reproduced with permission of John Wiley & Sons, Inc.)

Ostracoderm *Cephalaspis lyelli* from the Early Devonian of Scotland.

Denver Museum of Nature & Science

Use these references to answer the questions below.

Gee, H. (1996). Fishy fragments tip the scales, Nature News Service, http://www.geology. ucdavis.edu/~cowen/HistoryofLife/Anatolepis.html. Or see Janvier, P. (1996). Fishy fragments tip the scales. Nature 383, 757–758.; doi:10.1038/383757a0, in your college library. http://www.nature.com/nature/journal/v383/n6603/abs/383757a0.html.

University of California Museum of Paleontology, (2005). Vertebrates: Fossil Record, http://www. ucmp.berkeley.edu/vertebrates/vertfr.html.

University of California Museum of Paleontology, (2005). Vertebrates: Systematics, Early Vertebrate Groups, http://www.ucmp.berkeley.edu/vertebrates/vertsy.html.

University of California Museum of Paleontology, (1997). Introduction to the Petromyzontiformes, Lampreys, http://www.ucmp.berkeley.edu/vertebrates/basalfish/petro.html.

Young, G.C., Karatajute-Talimaa, V.N., & Smith, M.M. (1996). A possible Late Cambrian vertebrate from Australia. Nature 383, 810-812. doi:10.1038/383810a0, in your college library. http://www.nature.com/nature/journal/v383/n6603/abs/383810a0.html.

16. How old are the oldest known fish fossils, according to the 1996 Henry Gee article and the 1996 Philippe Janvier article? _____

17. Where were they collected? _____

18. What are ostracoderms? _____

19. What material composed the internal skeleton of the ostracoderms?

    _____

20. Why is the internal skeleton of ostracoderms poorly preserved or not preserved?

    _____

21. Why do you think that fish once had external skeletal armor?

    _____

22. Name two jawless fish living today.

    _____   _____

23. Describe the mouth of the lamprey.

    _____

A more-recent discovery has led to even older fossil fish. Use these references to answer the questions below.

Oldest fossil fish caught. (1999). BBC News, http://news.bbc.co.uk/1/hi/sci/tech/504776.stm.

Shu D. –G., Conway Morris, S., Han, J., et al. (2003).Head and backbone of the Early Cambrian vertebrate *Haikouichthys*. Nature 421, 526–529, http://www.nature.com/nature/journal/v421/n6922/abs/nature01264.html.

24. How old are these vertebrate fossils? (Give period name.)

    _____

25. Where were they discovered? _____

26. What are the genera of the two species of fish?

    _____

    _____

27. What features do these fossils have that leads paleontologists to conclude that they are fish? _____

28. What do these creatures *lack* that most vertebrates have?

    _____

29. These fish are from what is known as the _____ fossil fauna, where thousands of exquisitely preserved fossils of soft-bodied creatures have been found.

See these articles about Ordovician fossil fish:

Janvier, P. (1997). Tree of Life Web Project, Arandaspida, http://tolweb.org/tree?group=Arandaspida&contgroup=Pteraspidomorphi.

Janvier P. (1997). Tree of Life Web Project, Astraspida, http://tolweb.org/tree?group=Astraspida&contgroup=Pteraspidomorphi.

Read the articles and answer the following questions.

30. Did they have jaws? _____

31. Did they have armor? _____

32. What surrounds the eyes? _____ (More on this below.)

33. Did they have gills? _____

34. Did they have scales? _____

35. Give the age (period name and dates) of the earliest fish from these two references.

_____

36. Where did *Astraspis* live? (List four states where they have been found).

_____

_____

THE EVOLUTION OF FISH WITH JAWS

Jawed fish are gnathostomes, as opposed to the ostracoderms, or jawless fish. The classification of the gnathostomes is shown in Box 12.2. (A simplified version of some of the traditional nomenclature is included in blue.)

Box 12.2 • Classification of the Gnathostomes

GNATHOSTOMES

•••••• Osteostraci (fossil armored jawless vertebrates called ostracoderms)

•••••• Gnathostomata (vertebrates with jaws)

••••••• Placodermi (placoderm fish: armored jawed vertebrates)

••••••• Chondrichthyes (sharks and rays)

••••••• Teleostomi (teleost fish)

•••••••• Acanthodii (spiny fish)

•••••••• Osteichthyes (bony fish)

••••••••••• Actinopterygii (ray-finned fish; most modern fish)

••••••••••• Sarcopterygii (lobe-finned fish and terrestrial vertebrates) [Class Sarcopterygii]

••••••••••••Coelacanthimorpha (coelacanths) [Order Coelacanthiformes]

••••••••••••Dipnoi (lung fish) [Subclass Dipnoi]

••••••••••••Onychodontiformes (extinct fish)

••••••••••••Porolepimorpha (extinct fish)

••••••••••••Osteolepimorpha (extinct lobe-finned fishes such as *Eusthenopteron*)

••••••••••••Rhizodontimorpha (extinct lobe-finned fishes)

••••••••••••Terrestrial vertebrates [Subclass Tetrapodomorpha]

Janvier, P., (1997). Gnathostomata, Jawed Vertebrates, Tree of Life Web Project, http://tolweb.org/tree?group=Gnathostomata&contgroup=Vertebrata. (Scroll down to see the diagram relating jaws (in red) to gill arches (in green) in this reference.)

Zimmer, C. (1996). Death From the Pleistocene Sky. Discover 17(3), http://discovermagazine.com/1996/mar/deathfromtheplei720.

37. What animal is shown in the diagram with jaws and gill arches in the Gnathostomata reference? (Hint: See the caption.)

_____

38. What does the Zimmer article propose as a reason for the development of jaws?

_____

PLACODERM FISH

Placoderm means "plate skin": *plac* means "plate" and *derm* means "skin." The placoderm fish appeared in the Silurian, reached their greatest diversity in the Devonian (a time known as the Age of Fishes), and became extinct at the end of the Devonian. They had no true teeth (although the jaws had sharp tusklike projections) and no preservable internal skeleton. The only structure made of bone was the external armor.

Denver Museum of Nature & Science

Dunkleosteus, an extinct (Devonian) marine placoderm fish found in Ohio. The skull is more than 1 m high and 1 m wide.

Use these references to answer the questions below.

American Museum of Natural History, Hall of Vertebrate Origins, http://www.amnh.org/exhibitions/permanent-exhibitions/fossil-halls/hall-of-vertebrate-origins.

American Museum of Natural History, Dunkleosteus, http://www.amnh.org/exhibitions/permanent-exhibitions/fossil-halls/hall-of-vertebrate-origins/dunkleosteus.

Anderson, P.S.L. & Westneat, M.W. (2007). Feeding mechanics and bite force modeling of the skull of *Dunkleosteus terrelli*, an ancient apex predator. Biology Letters 3, 76-79, http://rsbl.royalsocietypublishing.org/content/3/1/77.full.

Cleveland Museum of Natural History, *Dunkleosteus terrelli*, http://www.cmnh.org/site/AtTheMuseum/OnExhibit/PermanentExhibits/Dunk.aspx.

University of California Museum of Paleontology, (2000). Introduction to the Placodermi: Extinct Armored Fishes with Jaws, http://www.ucmp.berkeley.edu/vertebrates/basalfish/placodermi.html.

39. In what formation were the *Dunkleosteus* fossils found? _____

40. How long was *Dunkleosteus*? _____ feet

41. How much did *Dunkleosteus* weigh? _____ tons

42. Was *Dunkleosteus* a predator or a scavenger? _____

43. What is suction feeding and how did it work? _____

_____

44. *Dunkleosteus* did not have teeth, so how did it bite? Describe the jaw and how it stayed sharp.

_____

45. How did its bite compare to other animals, including fish, mammals, alligators, and dinosaurs? _____

46. What did *Dunkleosteus* eat? _____

Look at the skull of the *Dunkleosteus* in the photo. Examine the eye. Notice the **sclerotic ring** (the bony structure in the eye socket for the protection of the eye). All birds have this sclerotic ring. It protects their eyes against rapid air pressure changes during flights. It is also used to help the eye focus on distant objects. Read more about the sclerotic ring here:

University of California, Museum of Paleontology, (2000). Eyes of Ichthyosaurs, http://www. ucmp.berkeley.edu/people/motani/ichthyo/eyes.html.

47. Why did swimming animals such as ichthyosaurs and *Dunkleosteus* need a sclerotic ring? _____

CHONDRICHTHYES: SHARKS

Chondrichthyes means "cartilage fish": *chondr* means "cartilage" and *ichthy* means "fish." Chondrichthyes are fish that have skeletons of cartilage instead of bone. This group includes the sharks, skates, rays, and others. The group appeared in the Ordovician and is still around today. Use the following references to learn about the largest shark that ever lived.

Florida Museum of Natural History, Ichthyology, *Megalodon*: Largest Shark that Ever Lived, http://www.flmnh.ufl.edu/fish/sharks/fossils/megalodon.html; also follow links on the right side of the page.

Florida Museum of Natural History, (2007). Educator's Guide for *Megalodon*: Largest Shark that Ever Lived, http://www.flmnh.ufl.edu/fish/sharks/fossils/megalodon-guide.pdf

University of California Museum of Paleontology, (1995). Introduction to the Chondrichthyes, http://www.ucmp.berkeley.edu/vertebrates/basalfish/chondrintro.html.

University of California Museum of Paleontology, (1995). Chondrichthyes: Fossil Record, http://www.ucmp.berkeley.edu/vertebrates/basalfish/chondrofr.html.

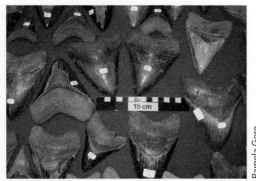

Fossilized teeth of *Carcharodon megalodon* from Georgia and South Carolina.

48. What is the largest shark that ever lived? _____

49. How long was it? _____

50. How does its size compare to that of *Dunkleosteus*? _____

51. When did it live? _____

52. The shark skeleton is composed of what? _____

53. What types of fossilized remains do we find of sharks? _____

54. *Megalodon* lived in a number of areas around the world, and its remains are found in several states in the United States. To which states might you go to collect *Megalodon* teeth? _____

OSTEICHTHYES: BONY FISHES

Osteichthyes means "bony fish": *osteo* means "bone" and *ichthys* means "fish." This group includes most modern fish, as well as a number of extinct fish. There are two major groups within the Osteichthyes: the **Actinopterygii** (actinopterygians) and the **Sarcopterygii** (sarcopterygians). The actinopterygians are "ray-finned" fish, with fins consisting of a web of skin supported by bony spines. (Most modern fish are of this type.) The sarcopterygians are "lobe-finned" fish, with bones and muscles in their fins. The lobe-finned fish appeared in the Devonian. The sarcopterygians are ancestral to the four-legged vertebrates (tetrapods). The sarcopterygians include the coelacanths and the dipnoi (or lungfish), along with *Eusthenopteron* and the tetrapods. Use these references to answer questions below and to learn more about the various types of bony fishes.

American Museum of Natural History, Hall of Vertebrate Origins, Coelacanth, http://www.amnh. org/exhibitions/permanent-exhibitions/fossil-halls/hall-of-vertebrate-origins/coelacanth.

Buchheim, H. P., (1998). A walk through time at Fossil Butte: Historical geology of the Green River Formation at Fossil Butte National Monument, http://nature.nps.gov/geology/paleontology/pub/ grd3_3/fobu1.htm.

Hamlin, J. F., (2013). The Fish Out of Time, Coelacanth Information, http://www.dinofish.com/.

Lundberg, J.G., (2006). Tree of Life Web Project, Actinopterygii: The ray-finned fishes, http:// tolweb.org/Actinopterygii/.

Tree of Life Web Project, (1995). Sarcopterygii: The Lobe-Finned Fishes and Terrestrial Vertebrates, http://tolweb.org/Sarcopterygii/.

University of California Museum of Paleontology, (1997). Introduction to the Actinopterygii: Fins and Bones, http://www.ucmp.berkeley.edu/vertebrates/actinopterygii/actinintro.html.

University of California Museum of Paleontology, (1997). Actinopterygii: Life History and Ecology, http://www.ucmp.berkeley.edu/vertebrates/actinopterygii/actinolh.html.

University of California Museum of Paleontology, Actinopterygii: More on Morphology, http:// www.ucmp.berkeley.edu/vertebrates/actinopterygii/actinomm.html.

University of California Museum of Paleontology, (1999). U.C. Berkeley Researchers Announce Discovery of Sulawesi Coelacanths, http://www.ucmp.berkeley.edu/vertebrates/coelacanth/ coelacanths.html; also follow links on left side of page.

University of California Museum of Paleontology, (2000). Introduction to the Sarcopterygii: from fins to legs, http://www.ucmp.berkeley.edu/vertebrates/sarco/sarcopterygii.html.

University of California Museum of Paleontology, (2000). Introduction to the Dipnoi: the lungfish, http://www.ucmp.berkeley.edu/vertebrates/sarco/dipnoi.html.

University of California Museum of Paleontology, (1999). Localities of the Eocene: The Green River Formation, http://www.ucmp.berkeley.edu/tertiary/eoc/greenriver.html.

Courtesy of Smithsonian Institution. Photo by Pamela Gore

Denver Museum of Nature & Science

Eusthenopteron was a sarcopterygian or lobe-finned fish. It is structurally similar to amphibians and is considered to be transitional to the amphibians. Escuminac Formation (Late Devonian). Quebec, Canada. About 25 cm long.

Another group of lobe-finned sarcopterygian fish invaded the sea and gave rise to the coelacanths (coelacanthimorpha). The coelacanths are considered to be living fossils because they were long-believed to be extinct, but one was caught in 1938 near Madagascar. More have been caught since.

Latimeria, a modern sarcopterygian coelacanth living near Madagascar. Note the similarity of the tail to that of *Eusthenopteron* fossils in the previous figure. The tail is very different from that of the ray-finned fishes. This example is about 2 m long. (From Levin, H., 2013, *The Earth Through Time* (10th edition), figure 12-75, p. 370. This material is reproduced with permission of John Wiley & Sons, Inc.)

55. Compare the shape of the tail of the living coelacanth to that of *Eusthenopteron.* Describe the similarities and differences.

   _____

   _____

56. Compare the tail of the coelacanth to that of the ray-finned fishes from the Green River Formation. In the boxes below, sketch both and describe the differences.

Tail of Coelacanth	Tail of ray-finned fish

   _____

57. Describe the function of the swim bladder. _____

58. Which type of fish was the most likely ancestor to the amphibians?

   _____

Some of the best-preserved fossil fish are in the Green River Formation (Eocene) in Wyoming.

Ray-finned fishes (actinopterygians) from the Green River Formation (Eocene), Wyoming. The largest fish is about 25 cm long.

59. What was the climate like in Wyoming during the Eocene?

    _____

60. What types of plants lived in Wyoming at that time?

    _____

61. Did these fish live in lakes or in the sea? _____

62. What type of rock contains these fish fossils? _____

63. Some living actinopterygians fish can walk on land and/or breathe air.

    a. List and describe them. _____

    b. Where do they live? _____

Mudskipper, *Periophthalmus novemradiatus*, a living actinopterygians fish (Southeast Asia), can use its muscular fins to walk out of the water, and it can breathe on land.

Pamela Gore

64. What do the fish that walk on land eat? _____

    _____

65. What do they do during times of periodic drying?

    _____

    _____

Go online to research either the air-breathing, walking "snakehead fish" from Asia, an invasive species that has appeared in the Potomac River and several other areas in the United States, *or* the "walking catfish" in Florida, another invasive species from Asia that walked away from a fish farm.

66. Write a paragraph telling whether these fish live in freshwater or in salt water, how they breathe, what they eat, how they move on land, how many eggs they lay, and why they are a problem.

    _____

    _____

    _____

    _____

    _____

    _____

    _____

    _____

THE TRANSITION TO LAND

During the Paleozoic Era, both plants and animals (invertebrates and vertebrates) made the transition from water to land. Plants made the transition first and provided nutritional support for animals to make the transition later. The wingless insects moved onto land, followed by the first amphibian-like vertebrates. Use the following references to answer the questions.

Klappenbach, L. Storming the Beaches: Vertebrates Transition from Water to Land: http://animals.about.com/od/amphibians/ss/landtowater.htm.

Klappenbach, L. A Brief History of Life on Earth: An Overview of Events That Have Shaped Life on Our Planet, http://animals.about.com/od/evolution/a/brief-history-of-life.htm.

University of California Museum of Paleontology, Understanding Evolution, Important Events in the History of Life, http://evolution.berkeley.edu/evolibrary/article/0_0_0/evo_13# (Hold cursor over time line.)

67. When did the first plants appear on the land? (Give period name.) _____

68. When did the first animals appear on the land? (Give period name.) _____

69. Which group of vertebrate animals was the first to appear on land?

    _____

70. Why did plants make the transition to land before the animals?

    _____

71. For organisms to make the transition from the water to the land, they must meet several requirements. List the challenges (or problems that must be solved) for vertebrates to make the transition to land. Also note how vertebrates solved each problem. List four challenges and solutions.

Challenge	Solution

A simplified classification of terrestrial vertebrates is shown in Box 12.3. A simplified version of the traditional nomenclature is included in blue.

Box 12.3   Simplified Classification of Terrestrial Vertebrates

TERRESTRIAL VERTEBRATES

•••••••••Sarcopterygii (lobe-finned fish and terrestrial vertebrates) [Class Sarcopterygii]

•••••••••••Terrestrial vertebrates

•••••••••••• Seymouriamorpha (*Seymouria*), commonly grouped with the amphibians

•••••••••••• *Ichthyostega*, commonly grouped with the *amphibians*

•••••••••••••Tetrapoda (four-legged terrestrial vertebrates with fingers and toes instead of fins)

•••••••••••••• Amphibia (living amphibians: frogs, salamanders, etc.)

•••••••••••••• Reptilomorpha

•••••••••••••••• Amniota (reptiles, mammals, birds, dinosaurs)

••••••••••••••••• Reptilia (reptiles, birds, dinosaurs)

••••••••••••••••• Synapsida (mammals and their extinct relatives)

Laurin, M. (2011). Tree of Life Web Project: Terrestrial Vertebrates. Stegocephalians: Tetrapods and other digit-bearing vertebrates, http://www.tolweb.org/Terrestrial_Vertebrates/.

Laurin, M. & Gauthier, J. A. (2012). Tree of Life Web Project: Amniota: Mammals, reptiles (turtles, lizards, *Sphenodon*, crocodiles, birds) and their extinct relatives, http://www.tolweb.org/Amniota/.

University of California Museum of Paleontology, Introduction to the Tetrapoda, The Four-Legged Vertebrates, http://www.ucmp.berkeley.edu/vertebrates/tetrapods/tetraintro.html.

University of California Museum of Paleontology, Tetrapods: Fossil Record, http://www.ucmp.berkeley.edu/vertebrates/tetrapods/tetrafr.html.

University of California Museum of Paleontology, Tetrapods: Systematics, http://www.ucmp.berkeley.edu/vertebrates/tetrapods/tetrasy.html.

The terrestrial vertebrates include the tetrapods (*tetra* means "four" and *pod* means "feet"). These are the four-legged animals including modern amphibians, reptiles, birds, and mammals. The earliest four-legged vertebrates on land (such as *Seymouria, Ichthyostega,* and others) were widely considered to be amphibians or amphibian-like because they inhabited watery habitats, but they are grouped separately from the tetrapods in these references because they retain some primitive fishlike characteristics and are not as closely related as was once thought. They are listed with the amphibians in most textbooks and other sources, however, so keep in mind that alternative classifications exist and that *Seymouria* and *Ichthyostega* are generally called amphibians and grouped with the tetrapods.

The land-dwelling vertebrates (amphibians, reptiles, birds, and mammals) have the same general types of bones in their skeletons. They have the radius, ulna, humerus, femur, tibia, fibula, ribs, vertebrae (bones of the spine), sternum, and phalanges. Research the names of the bones that make up a vertebrate skeleton online.

Skeletal differences between groups of land vertebrates come from the modification of existing bones to perform a special function. Types of modifications include fusing several bones into one and elongating a bone. You can tell a great deal about the life processes of an animal by looking at the modifications of its skeleton.

72. Draw a sketch of a skeleton and label the radius, ulna, humerus, femur, tibia, fibula, ribs, vertebrae (bones of the spine), sternum, and phalanges.

AMPHIBIANS AND THE EARLY TERRESTRIAL VERTEBRATES

Amphibians were the first land-dwelling vertebrates. Today, most adult amphibians live on land and breathe air, but they lay their eggs in the water. Most young amphibians live in the water and are fishlike (tadpoles, for example).

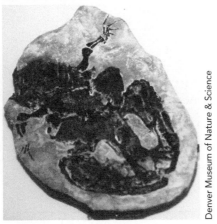

A. Modern salamander

Pamela Gore

B. *Saurerpeton obtusum,* Linton Formation, Middle Pennsylvanian, Jefferson County, Ohio.

Denver Museum of Nature & Science

C. A salamander ancestor, *Amphibamus lyelli,* Linton Formation, Middle Pennsylvanian, Jefferson County, Ohio.

Denver Museum of Nature & Science

Use these references to answer the following questions.

University of California Museum of Paleontology, (1995). Amphibia: Fossil Record, http://www.ucmp.berkeley.edu/vertebrates/tetrapods/amphibfr.html.

University of California Museum of Paleontology, (1995). Amphibia: Life History and Ecology, http://www.ucmp.berkeley.edu/vertebrates/tetrapods/amphiblh.html.

73. List the major characteristics of amphibians.

_____

74. When did the amphibians first appear? (Give period name.) _____

75. What were the ancestors of the amphibians? _____

The skeleton of *Ichthyostega,* an amphibian-like vertebrate. (From Levin, H., (2013), *The Earth Through Time,* (10th edition), figure 12-78, p. 372. This material is reproduced with permission of John Wiley & Sons, Inc.)

Ichthyostega, an amphibian-like vertebrate that appeared in the Late Devonian, retained many of the features of its fish ancestors. Use all of these references to answer the following questions on *Ichthyostega*.

Lauren, M. (2011). Tree of Life Web Project, Terrestrial Vertebrates: Stegocephalians: Tetrapods and other digit-bearing vertebrates, http://www.tolweb.org/Terrestrial_Vertebrates/.

Clack, J. A., (2006). Tree of Life Web Project, *Acanthostega gunneri*, http://tolweb.org/tree?group=Acanthostega&contgroup=Terrestrial_Vertebrates.

Clack, J. A., (1997). Tree of Life Web Project, The Definition of the Taxon Tetrapoda, http://tolweb.org/accessory/Definition_of_the_Taxon_Tetrapoda?acc_id=503.

Clack, J. A., (2006). Tree of Life Web Project, *Ichthyostega*, http://tolweb.org/Ichthyostega/.

Murphy, D., (2006). Devonian Times, *Ichthyostega* spp., http://www.devoniantimes.org/Order/re-ichthyostega.html.

Murphy, D., (2006). Devonian Times, Recent Findings: Prologue - Fish Out of Water, http://www.devoniantimes.org/Order/old-order.html.

Murphy, D., (2006). Devonian Times, Recent Findings: Fishes with Legs, http://www.devoniantimes.org/Order/new-order.html.

Zimmer, C., (1995). Coming onto the Land, Discover Magazine, http://discovermagazine.com/1995/jun/comingontothelan523.

76. What was Alfred Romer's scenario for the evolution of tetrapods in the 1940s and 1950s? (See the Zimmer article.)

_____

_____

_____

77. What are the fishlike characteristics of *Ichthyostega*?

_____

78. What are the amphibian-like characteristics of *Ichthyostega*?

_____

79. Describe the limbs of *Ichthyostega*.

_____

80. How does the bone structure of the limbs compare with that of the lobe-finned fishes (rhipidistians)?

_____

81. What are the new findings that alter our ideas of tetrapod evolution?

_____

_____

82. What is the significance of *Acanthostega*?

_____

83. Did *Ichthyostega* and other mid-Paleozoic amphibians inhabit freshwater or marine (salt) water?

_____

For 50 million years, from the Late Devonian to the Middle Carboniferous, amphibians were the only vertebrates to inhabit the land. Some adult amphibians reverted to an aquatic mode of life, and others retained a terrestrial lifestyle.

Seymouria was a land-dwelling amphibian from the Lower Permian of Texas. A primitive amphibian similar to *Seymouria* was probably ancestral to the reptiles.

Seymouria, a fossil amphibian, was a land dweller (Lower Permian), Texas. Note the stout limbs and the short body and tail. The specimen is a little less than 3 feet long. A primitive amphibian similar to *Seymouria* was probably ancestral to the reptiles.

Neopteroplax was an aquatic amphibian. *Eryops* was probably semiaquatic. Features suggesting an aquatic lifestyle include a flattened body and skull, reduced limbs, and a slender snakelike body.

Skull of *Neopteroplax,* an aquatic amphibian (Late Carboniferous), Ohio.

The large amphibian, *Eryops* (Permian).

A large number of well-preserved amphibian trackways were collected from an abandoned coal strip mine in Walker County, Alabama, between 1999 and 2004. Use the references to answer the questions.

Birmingham Paleontological Society, (2000). January 23, 2000 - Pennsylvanian Fossils, Walker County, AL, http://bps-al.org/trips/20000123.html.

Birmingham Paleontological Society, (2000). May 28, 2000 - Pennsylvanian Fossils, Walker County, AL, http://bps-al.org/trips/20000528.html.

Buta, R., (2005). Photographic Trackway Database, http://bama.ua.edu/~rbuta/monograph/database/database.html.Martin, A. J. & Pyenson, N. D., (2005). Behavioral Significance of Vertebrate Trace Fossils from the Union Chapel Site. In Buta, R. J., Rindsberg, A. K. & Kopaska-Merkel, D. C., eds. Pennsylvanian Footprints in the Black Warrior Basin of Alabama. Alabama Paleontological Society Monograph no 1, http://www.envs.emory.edu/faculty/MARTIN/ResearchDocs/Martin&Pyenson2005.pdf.

Sever, M., (2003). Mine Reclamation Threatens Tracksite, Geotimes, http://www.geotimes.org/oct03/NN_mine.html#.

Georgia Mineral Society, (2013). GMS Field Trip. Union Chapel Mine, Jasper, AL, http://www.gamineral.org/ft/2013/ft201302.html.

84. What is the age of the amphibian trackways (period name)? _____

85. What sorts of trace fossils are present? _____

86. What types of organisms made them? _____

87. How big were the amphibians? _____

REPTILES

Complete colonization of land was achieved by the reptiles, which could lay eggs on land. Egg-laying, however, is not easy to identify in the fossil record. There are two characteristics of the skull that can be used to distinguish reptiles from amphibians:

- The reptile skull is high and narrow, whereas the amphibian skull is low and broad.
- The roof of the reptile's mouth is arched and has small openings. The roof of the amphibian's mouth is flat and has large openings.

Modern lizard, a type of reptile. (About 12 cm long.)

Pamela Gore

Courtesy of Smithsonian Institution. Photo by Pamela Gore

Dicynodon, a plant-eating reptile (Late Permian, 250–230 million years ago), Cape Province, South Africa.

Answer the following questions using this reference. (Be sure to scroll down.)

Laurin, M. & Gauthier, A., (2012). Tree of Life Web Project: Amniota, http://tolweb.org/tree?group=Amniota&contgroup=Terrestrial_Vertebrates.

88. What are amniotes? _____

89. What are the oldest amniotes currently known (give genus) and where were they found?

    Genus _____

    Locality _____

90. What is the age of the oldest known amniotes? (Give period name.)

    _____

91. The amniotes diverged into two lines. List them below. Also list what they include or gave rise to.

    _____, which includes or gave rise to _____

    _____, which includes or gave rise to _____.

92. Describe the amniotic egg.

    _____

    _____

See the references to learn about temporal fenestration (holes in the side of the skull).

Olsen, P. E., (2006). Lecture 9, The end of the Permian and the Major Groups of Tetrapods, Lamont-Doherty Earth Observatory, Columbia University, http://www.ldeo.columbia.edu/dees/courses/v1001/permtrias8.html.

Tree of Life Web Project, (1996). Temporal Fenestration and the Classification of Amniotes, http://tolweb.org/notes/?note_id=463.

Laurin, M., (1996). University of California Museum of Paleontology: Anapsida: More on Morphology, http://www.ucmp.berkeley.edu/anapsids/anapsidamm.html.

93. List the animals that belong to each group:

   Anapsida (no holes) _____

   Diapsida (two holes) _____

   Euryapsida (upper hole only) _____

   Synapsida (lower hole only) _____

Diagram showing four vertebrate skull types. (From Levin, H., 2003, *The Earth Through Time* (7th edition), p. 367. This material is reproduced with permission of John Wiley & Sons, Inc.)

SYNAPSIDS: MAMMALS AND THEIR EXTINCT RELATIVES

Synapsids include mammals and their extinct amniote relatives that are more closely related to mammals than to reptiles. (Some earlier books referred to these extinct amniotes as "mammal-like reptiles," but they are more mammal-like than reptile-like.) For more information see the following reference.

Laurin, M. & Reisz, R. R., (2011). Tree of Life Web Project, Synapsida: Mammals and their extinct relatives, http://tolweb.org/Synapsida/.

In the Permian, the synapsids were the dominant terrestrial vertebrate. The synapsids gave rise to the mammal-like forms and then to the mammals. The best known group of Permian synapsids was the **pelycosaurs**, several of which had sails on their backs, supported by spines from their vertebrae. Two well-known pelycosaurs that evolved their sails independently were *Dimetrodon* and *Edaphosaurus*.

94. What are the major characteristics of the synapsids?

   _____

   _____

   _____

Permian pelycosaur, *Dimetrodon*.

Permian pelycosaur, *Edaphosaurus.*

See the following references for information on *Dimetrodon, Edaphosaurus* and their relatives.

American Museum of Natural History, *Dimetrodon*, http://www.amnh.org/exhibitions/permanent-exhibitions/fossil-halls/hall-of-primitive-mammals/dimetrodon.

Palaeos: Synapsida, http://palaeos.com/vertebrates/synapsida/index.html.

Palaeos: Synapsida: Overview, http://palaeos.com/vertebrates/synapsida/synapsida.html.

Palaeos: Synapsida, Pelycosauria, http://palaeos.com/vertebrates/synapsida/pelycosauria.html.

Palaeos: Edaphosauridae: Lupeosaurus, Glaucosaurus & Edaphosaurus, http://palaeos.com/vertebrates/synapsida/edaphosauridae2.html.

Palaeos: Edaphosauridae, http://palaeos.com/vertebrates/synapsida/edaphosauridae.html#Edaphosauridae.

Palaeos: Synapsida: Sphenacodontia, http://palaeos.com/vertebrates/synapsida/sphenacodontidae.html.

Olsen, P. E., (2006). Dinosaurs and the History of Life, Lecture 9, The end of the Permian and the Major Groups of Tetrapods, Lamont-Doherty Earth Observatory, Columbia University, http://www.ldeo.columbia.edu/dees/courses/v1001/permtrias8.html.

95. Compare the pictures of *Dimetrodon* and *Edaphosaurus*. Which genus has the larger head, with respect to body size? _____

96. Note the differentiated teeth of *Dimetrodon*. What did it probably eat?

_____

97. Note the blunt, almost peglike teeth of *Edaphosaurus*. What did it probably eat?

_____

98. What were two possible functions of the sails on the backs of the pelycosaurs?

_____

_____

99. How do we know that pelycosaurs were synapsids? _____

_____

100. When did the early true mammals first appear? _____

101. Write a few sentences to compare and contrast the mammal skulls in the photos below. Pay particular attention to the differences in dentition (teeth), elongation of the face, size and position of the eyes (i.e., are the eyes on the front or the side of the face?), and overall shape. Discuss the dolphin, coyote, mountain lion, cat, and horse. Put in the form of a list so there is room to write something about each.

102. Examine the photo of a cloven hoof. How does this differ from the hoof of a horse? _____

103. Is there an even or odd number of toes? _____

Mammalian Skulls

Dolphin.

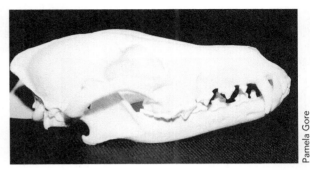

Coyote.

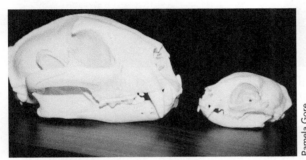

Mountain lion and cat.

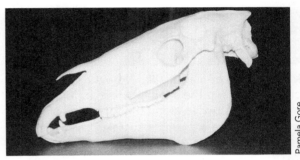

Horse.

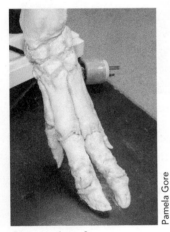

Cloven hoof.

THE PRIMATES

Primates are a group of mammals with five digits, which is a primitive, structurally generalized characteristic. Primates have not developed specialized features such as hooves, horns, antlers, or trunks. Some of the main characteristics of the primates include:

- Enlarged brain
- Shortened and flattened face (compared with other mammals; compare photos of skulls)
- Opposable thumb
- Forearm mobility, permitting rotation of the ulna and radius, so that the hand can be turned
- Modifications of the thorax, allowing upright posture
- Forelimbs and hind limbs that have different form and function
- Eyes that are close-set and positioned toward front of the face, allowing binocular stereoscopic vision and the ability to judge distance

These adaptations might have been beneficial to an arboreal (tree-dwelling) existence or to catching prey.

Primate Skulls

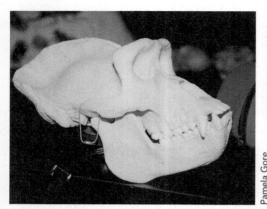

Male gorilla, *genus Gorilla*. Note the prominent brow ridge, and the saggital crest, a ridge of bone running lengthwise along the top of the skull.

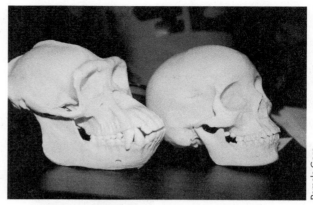

Chimp versus human. *Left,* chimpanzee, *Pan troglodytes; Right, Homo sapiens.*

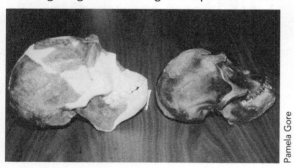

Left, Homo erectus; Right, Australopithecus.

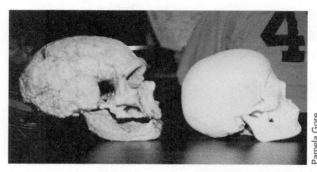

Left, Homo neanderthalensis; Right, Homo sapiens.

104. Write a paragraph comparing and contrasting primate skulls in terms of size, flattening of the face, presence or absence of brow ridge, dentition (teeth), robustness of jaw, and presence of a sagittal crest (or front-to-back ridge across the top of the skull), discussing the gorilla, chimp (*Pan troglodytes*), *Homo erectus, Australopithicus, Homo neanderthalensis*, and *Homo sapiens*.

_____

_____

_____

_____

_____

105. Which skull is larger: *Homo sapiens* or *Homo neanderthalensis?* _____

106. What do you think this means? _____

_____

_____

_____

_____

_____

Eon	Era	Period	Epoch	Date (Millions of years before present)
Phanerozoic	Cenozoic	Quaternary	Holocene	
			Pleistocene	
		Neogene	Pliocene	
			Miocene	
		Paleogene	Oligocene	
			Eocene	
			Paleocene	
	Mesozoic	Cretaceous	End of Mesozoic Era 66 million years ago	
		Jurassic		
		Triassic		
	Paleozoic	Permian	End of Paleozoic Era 252 million years ago	
		Carboniferous	Pennsylvanian	
			Mississippian	
		Devonian		
		Silurian		
		Ordovician		
		Cambrian		
Proterozoic	Precambrian (87% of geologic time scale)		End of Precambrian 541 million years ago	
Archean			End of Archean Eon 2.5 billion years ago	
Hadean (informal)			End of Hadean 4 billion years ago	

Ages of this chart are according to "The Geologic Time Scale 2012", Gradstein, F.M., Ogg, J.G., Schmitz, M.D., and Ogg, G.M., Elsevier, 2 Volume set. Colors are according to the Commission for the Geological Map of the World.